SCORECASTING

*The Hidden Influences
Behind How Sports Are Played
and Games Are Won*

Tobias J. Moskowitz and L. Jon Wertheim

THREE RIVERS PRESS
NEW YORK

Copyright © 2011 by Tobias J. Moskowitz and L. Jon Wertheim

All rights reserved.
Published in the United States by Three Rivers Press, an imprint of the
Crown Publishing Group, a division of Random House, Inc., New York.
www.crownpublishing.com

Three Rivers Press and the Tugboat design are registered trademarks of
Random House, Inc.

Originally published in hardcover in the United States by Crown Archetype,
an imprint of the Crown Publishing Group, a division of
Random House, Inc., New York, in 2011.

Library of Congress Cataloging-in-Publication Data
Moskowitz, Tobias J. (Tobias Jacob), 1971–
Scorecasting: The hidden influences behind how sports are played
and games are won / Tobias J. Moskowitz and L. Jon Wertheim.
         p.   cm.
1. Sports—Miscellanea.   2. Sports—Problems, exercises, etc.
I. Wertheim, L. Jon.   II. Title.
GV707.M665 2011
796—dc22
2010035463

ISBN 978-0-307-59180-7
eISBN 978-0-307-59181-4

Printed in the United States of America

Book design by R. Bull
Cover design by Kyle Kolker

10 9 8 7

First Paperback Edition

Additional Praise for **SCORECASTING**

"The closest thing to *Freakonomics* I've seen since the original. A rare combination of terrific storytelling and unconventional thinking. I love this book."

—Steven D. Levitt, bestselling coauthor of
Freakonomics and *Superfreakonomics*

"A counterintuitive, innovative, unexpected handbook for sports fans interested in the truths that underpin our favorite games. With their lively minds and prose, Moskowitz and Wertheim will change the way you think about and watch sports. Not just for stats nerds, *Scorecasting* enlightens and entertains. I wish I had thought of it!"

—Jeremy Schaap, ESPN reporter,
author of *Cinderella Man*

"(Sports + numbers) × great writing = winning formula. A must-read for all couch analysts."

—Richard Thaler, professor of behavioral science and
economics, bestselling author of *Nudge*

"*Scorecasting* is both scholarly and entertaining, a rare double. It gets beyond the clichéd narratives and tried-but-not-necessarily-true assumptions to reveal significant and fascinating truths about sports."

—Bob Costas

"*Scorecasting* will change the way you watch sports, but don't start reading it during a game; you're liable to get lost in it and miss the action. I'm not giving anything away, because you'll want to read exactly how they arrived at their conclusions."

—Allen Barra, *Newark Star Ledger*

"A must-read for all aficionados of sports or economics . . . The brilliance of *Scorecasting* lies in its ability to transform even the most know-it-all sports fan back into a bright-eyed prepubescent . . . with an insatiable curiosity and thirst for learning everything about [sports]."

—*Deseret News*

"*Scorecasting* is a book that sports fans should take to their upcoming tailgates. In one fell swoop, it shatters many of the most cherished athletic clichés with hard data and headstrong argument . . . Simultaneously shocking and sensible."

—*Washington Times*

To our wives, to our kids . . . and to our parents for driving us between West Lafayette and Bloomington all those years

CONTENTS

INTRODUCTION

It was the summer of 1984 in Ortonville, Michigan, a lakeside blip on the map somewhere between Detroit and Flint. The second session of Camp Young Judaea—province to a few hundred kids from the American heartland—was under way, and Bunk Seven fielded a formidable softball team.

There was one problem. In keeping with the camp's themes of community and democracy and egalitarianism and the like, the rules dictated that every member of the bunk was required to bat and play the field. Although eight members of Bunk Seven ranged from capable to exceptional softball players, the ninth was, in a word, tragic. One poor kid from Iowa whose gangly body resembled a map of Chile—we'll call him Ari, thus sparing anyone potential embarrassment—was a thoroughly pleasant bunkmate, armed with a vast repertoire of dirty jokes and a cache of contraband candy. Unfortunately, Ari was sensationally nonathletic. Forget catching a ball. Asking him to drink his "bug juice" from a straw would mean confronting the outer limits of his physical coordination. Robert Redford was starring in *The Natural* that summer, and here, on another baseball diamond, was the Unnatural.

Not surprisingly, when Bunk Seven took the field, Ari was

dispatched to the hinterlands of right field, on the fringes of the volleyball court, the position where, the conventional thinking went, he was least likely to interface with a batted ball. The games took on a familiar rhythm. Bunk Seven would seize an early lead. Eventually, Ari would come to the plate. He would stand awkwardly, grip the bat improperly, and hit nothing but air molecules with three swings. Glimpsing Ari's ineptitude, the opposing team would quickly deduce that he was the weak link. When it was their turn to bat, they would direct every ball to right field. Without fail, balls hit to that area would land over, under, or next to Ari—anywhere but in the webbing of his borrowed glove. Eventually he'd gather the ball and, with all those ungovernable limbs going in opposite directions, make a directionless toss. The other team would score many runs. Bunk Seven would lose.

A few weeks into the summer, the Bunk Seven brain trust seized on an idea: If Ari played catcher instead of right field, he might be less of a liability. On its face, the plan was counterintuitive. With Ari behind home plate, his clumsiness would be on full display, starting with the first pitch, and he'd figure prominently in the game, touching the ball on almost every play.

But there was no base stealing allowed, so Ari's woeful throwing wasn't a factor. He might drop the odd pop-up, but at least the ball would be in foul territory and the batter wouldn't advance around the bases the way he did when Ari dropped balls in right field. With the eight capable players in the field, Bunk Seven didn't let too many runners reach base. On the rare occasions when a runner might try to score, there was usually sufficient time for the pitcher or first baseman to cover the plate, gently relieving Ari of his duties—something that couldn't be done as easily on a ball batted to right field.

There was a more subtle, unforeseen benefit as well. On pitches that weren't hit, it took Ari an unholy amount of time to gather the ball and throw it back to the pitcher. This slowed the game's pace considerably. The pre-bar-mitzvah-aged attention span being what it is, the opposing team began swinging at bad pitches, if only to bypass the agony of waiting for Ari to retrieve the ball. And Bunk Seven's pitcher started tossing worse pitches as a result.

Ari never perfected the fine art of hitting, and eventually he ran out of contraband Skittles. But once he was positioned behind home plate, Bunk Seven didn't lose another softball game the rest of the summer.

■ ■ ■

For two members of Bunk Seven—a pair of sports-crazed 12-year-olds from Indiana, one named Moskowitz and the other Wertheim—this was instructive. The textbook strategy was to conceal your least competent player in right field and then hope to hell no balls were hit his way. But says who? By challenging the prevailing wisdom and experimenting with an alternative, we were able to improve the team and win more games.

We've been friends ever since, bound in no small part by a mutual love of sports. Now, a quarter century later—with one of us a University of Chicago finance professor and the other a writer at *Sports Illustrated*—we're trying to confront conventional sports wisdom again. The concepts might be slightly more advanced and the underlying analysis more complex, but in the forthcoming pages of *Scorecasting,* we're essentially replicating what we did on the camp softball field. Is it really preferable to punt on fourth down rather than go for it? To keep feeding the teammate with the hot hand? To try to achieve the highest available spot in the draft? Is there an *I* in *team?* Does defense truly win championships?

As for the sports truisms we accept as articles of faith, what's driving them? We *know,* for instance, that home teams in sports—in all sports, at any level, at any time in history—win the majority of the games. But is it simply because of rabid crowd support? Or is something else going on? As lifelong Cubs fans, we know all too well that without putting too fine a point on it, our team sucks. But is it simply because the Cubs are unlucky, somehow cursed by the baseball deities and/or an aggrieved goat? Or is there a more rational explanation?

Even though sports are treated as a diversion and ignored by highbrow types, they are imbued with tremendous power to explain human behavior more generally. The notion that "sports are a metaphor for life" has hardened into a cliché. We try to "be like Mike," to "go for the gold," to "just do it," to "cross the goal line," to "hit the home run."

The inverse is true, too, though. Life, one might say, is a microcosm for sports. Athletes and coaches may perform superhuman feats, but they're subject to standard rules of human behavior and economics just like the rest of us. We'll contend that an NFL coach's decision to punt on fourth down is not unlike a mutual fund manager's decision to buy or sell a stock or your decision to order meatloaf rather than the special of the day off a diner menu. We'll try to demonstrate that Tiger Woods assesses his putts the same way effective dieters persuade themselves to lose weight—and makes the same golfing mistakes you and I do. We'll explain how referees' decision-making resembles parents deciding whether to vaccinate their kids and why this means that officials don't always follow the rule book. We'll find out how we, as fans, view our favorite teams much the same way we look at our retirement portfolios, suffering from the same cognitive biases. As in life, much of what goes on in sports can be explained by incentives, fears, and a desire for approval. You just have to know where to look . . . and it helps if you have data to prove it.

Many of the issues we explore might seem unrelated and, in many cases, reach far beyond sports, but they are all held together by a common thread of insight that remains hidden from our immediate view. Exploring the hidden side of sports reveals the following:

- *That which is* recognizable or apparent *is often given too much credit, whereas the real answer often lies concealed.*
- *Incentives are powerful motivators and predictors of how athletes, coaches, owners, and fans behave—sometimes with undesirable consequences.*

- Human biases and behavior *play a pivotal role in almost every aspect of life, and sports are no exception.*
- *The* role of luck *is underappreciated and often misunderstood.*

These themes are present in *every* sport. The hidden influences in the National Football League are equally present in the National Basketball Association, or Major League Baseball, or soccer worldwide. The presence of these factors across many sports highlights how powerful and influential their effects are.

We're expecting that many of the statements and claims we'll make in the following chapters will be debated and challenged. If so, we have done our job. The goal of *Scorecasting* is not to tell you *what* to think about sports but rather *how* to think about sports a little differently. Ambitiously, we hope this book will be the equivalent of a 60-inch LCD, enabling you to see the next game a little more clearly than you might have before.

We may even settle a few bar fights. With any luck, we'll start a few, too.

WHISTLE SWALLOWING

Why fans and leagues
want officials to miss calls

If you don't have at least *some* sympathy for sports officials, consult your cardiologist immediately. It's not just that refs, umps, and linesmen take heaps of abuse. It's the myths and misconceptions. Fans are rarely so deluded as to suggest that they could match the throwing arm of Peyton Manning or defend Kobe Bryant or return Roger Federer's serve, but somehow every fan with a ticket or a flat-screen television is convinced he could call a game as well as the schmo (or worse) wearing the zebra-striped shirt.

This ignores the reality that officials are accurate—uncannily so—in their calls. It ignores the reality that much like the best athletes, they've devoted years of training to their craft, developed a vast range of skills and experiences, and made it through a seemingly endless winnowing process to get to the highest level. It also ignores the reality that most referees aren't lucky sports fans who were handed a whistle; they tend to be driven, and smart, and successful in their other careers as well.

Consider, for instance, Mike Carey. The son of a San Diego doctor, Carey was a college football player of some distinction until his senior year, when he injured his foot in a game. Any ambitions of playing in the NFL were shot, but that was okay. He

Don /
Mike Carey

graduated with a degree in biology from Santa Clara University and, an incurable tinkerer, founded a company, Seirus Innovation, that manufactures skiing and snowboarding accessories. Carey even owns a number of patents, including Cat Tracks, a device that slips over a ski boot to increase traction.

In his first year out of college, though, Carey realized that he had a knack for overseeing football games. Part of it was an ability to make the right call, but he also had a referee's intuition, a sixth sense for the rhythm and timing of a game. Plus, he cut a naturally authoritative figure. Just as a pro football player would, he showed devotion to the craft, working his way up from local Pop Warner games to high school to Division I college games to the NFL, where his older brother, Don, was already working as a back judge. Carey reached the pinnacle of his officiating career when he was selected as referee for Super Bowl XLII, the first African-American referee assigned to work the biggest event on the American sports calendar. (Don Carey worked as a back judge for Super Bowl XXXVII.)

Played on February 3, 2008, Super Bowl XLII was a football game that doubled as a four-quarter passion play. Heavily favored and undefeated on the season, the New England Patriots clung to a 14–10 lead over the New York Giants late in the fourth quarter. A defensive stop and the Patriots would become the first NFL team since the 1972 Miami Dolphins to go through an entire season undefeated—and the first team to go 19–0.

As the Giants executed their final drive, with barely more than a minute remaining, they were consigned to third down and five from their own 44-yard line. Eli Manning, the Giants' quarterback, took the snap and scrambled and slalomed in the face of a fierce Patriots pass rush, as if inventing a new dance step. He ducked, jived, spun, and barely escaped the clutches of New England's defensive line, displaying the footwork of Arthur Murray and the cool of Arthur Fonzarelli.

Finally, in one fluid motion, Manning adjusted, planted a foot, squared himself, and slung the ball to the middle of the field. His target was David Tyree. It was surprising to many that Tyree was even on the field. Usually a special teams player, he had caught only four passes all season and dropped a half dozen balls during the Friday practice before the game. ("Forget about it," Manning had said to him consolingly. "You're a gamer.") Compounding matters, Tyree was defended by Rodney Harrison, New England's superb All-Pro strong safety.

As Manning scrambled, Tyree, who had run a post pattern, stopped, and then loitered in the middle of the field, realizing that his quarterback was still looking for an open receiver. As the ball approached, Tyree jumped, reaching back until he was nearly parallel to the field. With one hand, he snatched the ball and pinned it against his helmet. Somehow, he held on to it for a 32-yard gain. Instead of a sack and a fourth down, Tyree and Manning had combined for an impossible "Velcro catch" that put the Giants on the Patriots' 24-yard line. Tyree would never catch another pass in the NFL, but it was a hell of a curtain call.

Four plays later, Manning would throw a short touchdown pass to Plaxico Burress and the Giants would pull off one of the great sports upsets, winning Super Bowl XLII, 17–14. It was "the Tyree pass" that everyone remembers. No less than Steve Sabol, the president of NFL Films and the sport's preeminent historian, called it "the greatest play in Super Bowl history."

The play was extraordinary, no doubt about it, but the officiating on it was quite ordinary. That is, the men in the striped uniforms and white caps did what they usually do at a crucial juncture: They declined to make what, to some, seemed like an obvious call. Spark up YouTube and watch "the Tyree play" again, paying close attention to what happens in the backfield. Before Manning makes his great escape, he is all but bear-hugged by a cluster of Patriots defenders—Richard Seymour and Adalius Thomas in particular—who had grasped fistfuls of the right side of his number 10 jersey. Manning's progress appeared to be stopped. Quarterbacks in far less peril have been determined to be "in the grasp," a determination made to protect quarterbacks that awards the defense with a sack when players grab—as opposed to actually tackle—the quarterback.

To that point, Mike Carey was having the game of his life. Everything had broken right. He had worked the Patriots-Giants game in the final week of the regular season (several weeks earlier), and so he had an especially well-honed sense for the two teams. "Just like athletes and teams, we were in the zone that night," he says, "both individually and as a crew."

More than two years later, Carey recalls the Tyree play vividly. He remembers being surprised that Manning hadn't used a hard count in an attempt to draw New England offside—that's how locked into the game he was. When the ball was snapped, Carey started on the left side of the field but then backpedaled and found an unobstructed view behind Manning. A few feet away from the play, alert and well positioned as usual, eyes lasering on the players, Carey appeared poised to declare that Manning was sacked. And then . . . nothing. It was a judgment call, and Carey's judgment was not to judge.

"Half a second longer and I would've had to [call him in the grasp]," says Carey. "If I stayed in my original position, I would have whistled it. Fortunately, I was mobile enough to see that he wasn't completely in the grasp. Yeah, I had a sense of 'Oh boy, I hope I made the right call.' And I think I did. . . . I'm glad I didn't blow it dead. I'd make the same call again, whether it was the last [drive] of the Super Bowl or the first play of the preseason."

Others aren't so sure. Reconsidering the play a year later, Tony Dungy, the former Indianapolis Colts coach and now an NBC commentator, remarked: "It should've been a sack. And, I'd never noticed this before, but if you watch Mike Carey, he almost blows the whistle. . . . With the game on the line, Mike gives the QB a chance to make a play in a Super Bowl. . . . I think in a regular season game he probably makes the call."* In other words, at

---

* It bears mention that Dungy made these remarks on an NBC broadcast while talking to his colleague Rodney Harrison, the defensive back who was covering Tyree on the play.

least according to Dungy, the most famous play in Super Bowl history might never have happened if the official had followed the rule book to the letter and made the call he would have made during the regular season.

It might have been a correct call. It might have been an incorrect call. But was it the *wrong* call? It sure didn't come off that way. Carey was not chided for "situational ethics" or "selective officiating" or "swallowing the whistle." Quite the contrary. He was widely hailed for his restraint, so much so that he was given a grade of A+ by his superiors. In the aftermath of the game, he appeared on talk shows and was even permitted by the NFL to grant interviews—including one to us as well as one to *Playboy*—about the play, a rarity for officials in most major sports leagues. It's hard to recall the NFL reacting more favorably to a single piece of officiating.

■ ■ ■

If this is surprising, it shouldn't be. It conforms to a sort of default mode of human behavior. People view acts of *omission*—the absence of an act—as far less intrusive or harmful than acts of *commission*—the committing of an act—even if the outcomes are the same or worse. Psychologists call this *omission bias,* and it expresses itself in a broad range of contexts.

In a well-known psychological experiment, the subjects were posed the following question: Imagine there have been several epidemics of a certain kind of flu that everyone contracts and that can be fatal to children under three years of age. About 10 out of every 10,000 children with this flu will die from it. A vaccine for the flu, which eliminates the chance of getting it, causes death in 5 out of every 10,000 children. Would you vaccinate your child?

On its face, it seems an easy call, right? You'd choose to do it because not vaccinating has twice the mortality rate as the vaccination. However, most parents in the survey opted *not* to vaccinate their children. Why? Because it *caused* 5 deaths per 10,000; never mind that without the vaccine, their children faced twice the risk of death from the flu. Those who would not per-mit vaccinations indicated that they would "feel responsible if anything happened because of [the] vaccine." The same parents tended to dismiss the notion that they would "feel responsible if anything had happened because I failed to vaccinate." In other words, many parents felt more responsible for a bad outcome if it followed their own actions than if it simply resulted from lack of action.

In other studies, subjects consistently view various actions *taken* as less moral than actions not taken—even when the results are the same or worse. Subjects, for instance, were asked to assess the following situation: John, a tennis player, has to face a tough opponent tomorrow in a decisive match. John knows his oppo-nent is allergic to a particular food. In the first scenario, John rec-ommends the food containing the allergen to hurt his unknowing opponent's performance. In the second, the opponent mistakenly orders the allergenic food, and John, knowing his opponent might get sick, says nothing. A majority of people judged that John's *action* of recommending the allergenic food was far more immoral than John's *inaction* of not informing the opponent of the aller-genic substance. But are they really different?

Think about how we act in our daily lives. Most of us probably would contend that telling a direct lie is worse than withholding the truth. Missing the opportunity to pick the right spouse is bad but not nearly as bad as actively choosing the wrong one. Declin-ing to eat healthy food may be a poor choice; eating junk food is worse. You might feel a small stab of regret over not raising your hand in class to give the correct answer, but raise your hand and provide the wrong answer and you feel much worse.

Psychologists have found that people view inaction as less causal, less blameworthy, and less harmful than action even when the outcomes are the same or worse. Doctors subscribe to this philosophy. The first principle imparted to all medical students is "Do no harm." It's not, pointedly, "Do some good." Our legal system draws a similar distinction, seldom assigning an affirmative *duty* to rescue. Submerge someone in water and you're in trouble.

Stand idly by while someone flails in the pool before drowning and—unless you're the lifeguard or a doctor—you won't be charged with failing to rescue that person.

In business, we see the same omission bias. When is a stockbroker in bigger trouble? When she neglects to buy a winning stock and, say, misses getting in on the Google IPO? Or when she invests in a dog, buying shares of Lehman Brothers with your retirement nest egg? Ask hedge fund managers and, at least in private, they'll confess that losing a client's money on a wrong pick gets them fired far more easily than missing out on the year's big winner. And they act accordingly.

In most large companies, managers are obsessed with avoiding actual errors rather than with missing opportunities. Errors of commission are often attributed to an individual, and responsibility is assigned. People rarely are held accountable for failing to act, though those errors can be just as costly. As Jeff Bezos, the founder of Amazon, put it during a 2009 management conference: "People overfocus on errors of commission. Companies overemphasize how expensive failure's going to be. Failure's not that expensive. . . . The big cost that most companies incur is much harder to notice, and those are errors of omission."

This same thinking extends to sports officials. When referees are trained and evaluated in the NBA, they are told that there are four basic kinds of calls: correct calls, incorrect calls, correct noncalls, and incorrect noncalls. The goal, of course, is to be correct on every call and noncall. But if you make a call, you'd better be right. "It's late in the game and, let's say, there's goaltending and you miss it. That's an incorrect noncall and that's bad," says Gary Benson, an NBA ref for 17 years. "But let's say it's late in the game and you call goaltending on a play and the replay shows it was an incorrect call. That's when you're in a *really* deep mess." *

Especially during crucial intervals, officials often take pains not to insinuate themselves into the game. In the NBA, there's an

---

* Ironically, Dallas Mavericks owner Mark Cuban earned one of his first (of many) fines when he disputed a late-game goaltending call that Benson refrained from making.

n directive: "When the game steps up, you step down."
ach as possible, you gotta let the players determine who
and loses," says Ted Bernhardt, another longtime NBA ref.
"It's one of the first things you learn on the job. The fans didn't
come to see you. They came to see the athletes."

It's a noble objective, but it expresses an unmistakable *bias,* and
one could argue that it is worse than the normal, random mistakes
officials make during a game. Random referee errors, though annoying, can't be predicted and tend to balance out over time, not
favoring one team over the other. With random errors, the system
can't be gamed. A systematic *bias* is different, conferring a clear
advantage (or disadvantage) on one type of player or team over
another and enabling us—to say nothing of savvy teams, players, coaches, executives, and, yes, gamblers—to predict who will
benefit from the officiating in which circumstances. As fans, sure,
we want games to be officiated accurately, but what we should
really want is for games to be officiated without bias. Yet that's not
the case.

■ ■ ■

Start with baseball. In 2007, Major League Baseball's website,
mlb.com, installed cameras in ballparks to track the location of
every pitch, accurate to within a centimeter, so that fans could
follow games on their handhelds, pitch by pitch. The data—called
Pitch f/x—track not only the location but also the speed, movement, and type of pitch. We used the data, containing nearly 2
million pitches and 1.15 million *called* pitches, for a different purpose: to evaluate the accuracy of umpires. First, the data reveal
that umpires are staggeringly accurate. On average, umpires make
erroneous calls only 14.4 percent of the time. That's impressive,
especially considering that the average pitch starts out at 92 mph,
crosses the plate at more than 85 mph, and usually has been gar-
nished with all sorts of spin and movement.

But those numbers change dramatically depending on the situation. Suppose a batter is facing a two-strike count; one more
called strike and he's out. Looking at all called pitches in baseball

over the last three years that are actually within the strike zone on two-strike counts (and removing full counts where there are two strikes and three balls on the batter), we observed that umpires make the correct call only 61 percent of the time. That is, umpires erroneously call these pitches balls 39 percent of the time. So on a two-strike count, umpires have more than twice their normal error rate—and in the batters' favor.

What about the reverse situation, when the batter has a three-ball count and the next pitch could result in a walk? Omission bias suggests that umpires will be more reluctant to call the fourth ball, which would give the batter first base. Looking at all pitches that are actually outside the strike zone, the normal error rate for an umpire is 12.2 percent. However, when there are three balls on the batter (excluding full counts), the umpire will erroneously call strikes on the same pitches 20 percent of the time.

In other words, rather than issue a walk or strikeout, umpires seem to want to prolong the at-bat and let the players determine the outcome. They do this even if it means making an incorrect call—or, at the very least, refraining from making a call they would make under less pressured circumstances.

The graph on page 16 plots the actual strike zone according to MLB rules, represented by the box outlined in black. Taking all called pitches, we plot the "empirical" strike zone based on calls the umpire is actually making in two-strike and three-ball counts. Using the Pitch f/x data, we track the location of every called pitch and define any pitch that is called a strike more than half the time to be within the empirical strike zone. The strike zone for two-strike counts is represented by the dashed lines, and for three-ball counts it is represented by the darker solid area.

The graph shows that the umpire's strike zone shrinks considerably when there are two strikes on the batter. Many pitches that are technically within the strike zone are not called strikes when that would result in a called third strike. Conversely, the umpire's strike zone expands significantly when there are three balls on the

ACTUAL STRIKE ZONE FOR THREE-BALL VERSUS TWO-STRIKE COUNTS

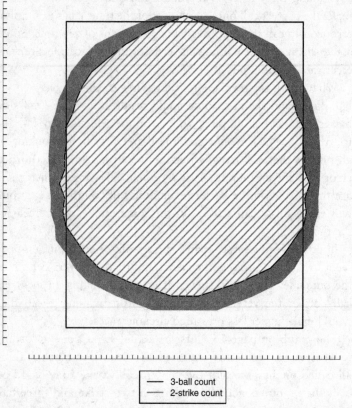

3-ball count
2-strike count

Box represents the rules-mandated strike zone. Tick marks represent a half inch.

batter, going so far as to include pitches that are more than several inches outside the strike zone. To give a sense of the difference, the strike zone on three-ball counts is 93 square inches larger than the strike zone on two-strike counts.*

---

* Notice that in both situations umpires tend to call high pitches strikes more often and call low pitches strikes far less often than the rules state that they should. This confirms what many baseball insiders have thought for years: MLB umpires have a high strike zone.

The omission bias should be strongest when making the right call would have a big influence on the game but missing the call would not. (Call what should be a ball a strike on a 3–0 pitch and, big deal, the count is only 3–1.) Keeping that in mind, look at the next graph. The strike zone is smallest when there are two strikes and no balls (count is 0–2) and largest when there are three balls and no strikes (count is 3–0).

ACTUAL STRIKE ZONE
FOR 0–2 AND 3–0 COUNTS

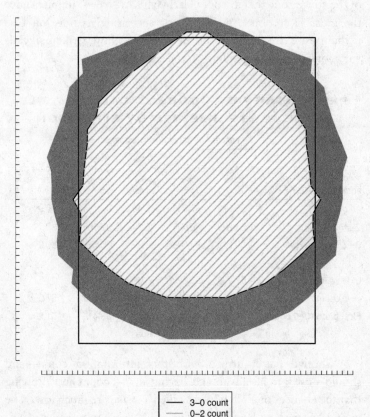

———— 3–0 count

———— 0–2 count

Box represents the rules-mandated strike zone. Tick marks represent a half inch.

The strike zone on 3–0 pitches is *188* square inches larger than it is on 0–2 counts. That's an astonishing difference, and it can't be a random error.

We also can look at the specific location of pitches. Even for obvious pitches, such as those in the dead center of the plate or those *waaay* outside the strike zone—which umpires rarely miss—the pitch will be called differently depending on the strike count. The umpire will make a bad call to prolong the at-bat even when the pitch is obvious. So what happens with the less obvious pitches? On the most ambiguous pitches, those just on or off the corners of the strike zone that are not clearly balls or strikes, umpires have the most discretion. And here, not surprisingly, omission bias is the most extreme. The table below shows how strike-ball calls vary considerably depending on the situation.

PERCENTAGE OF CORRECT CALLS OF MLB HOME PLATE UMPIRES BY SITUATION

	PERCENT OF CALLS THAT ARE CORRECT				
PITCH IS ACTUALLY:	In strike zone	Outside strike zone	Dead center of strike zone	Way outside strike zone	PERCENT OF STRIKES CALLED ON THE CORNERS
All counts	80.2%	87.8%	98.4%	92.1%	49.9%
2-strike counts	61.3%	93.5%	88.7%	97.0%	38.2%
3-ball counts	89.0%	84.0%	99.9%	86.5%	60.0%
0–2 counts	57.7%	93.8%	85.0%	98.4%	31.5%
3–0 counts	93.1%	80.0%	100.0%	82.0%	67.6%
First pitch (0–0 count)	84.9%	90.1%	98.9%	83.4%	51.2%

A shrewd batter armed with this information could—and should—use it to his advantage. Facing an 0–2 count and knowing that the chances of a pitch being called a strike are much lower, he would be smart to be conservative in his decision to swing. Conversely, on a 3–0 count, the umpire is much more likely to call a strike, so the batter may be better off swinging more freely.

From Little League all the way up to the Major Leagues, managers, coaches, and hitting experts all encourage players to "take the pitch" on 3–0. The thinking, presumably, is that the batter is so close to a walk, why blow it? But considering the home plate umpire's omission bias, statistics suggest that batters might be better off swinging, because they're probably conceding a strike otherwise. And typically, a pitcher facing a 3–0 count conservatively throws a fastball down the middle of the plate to avoid a walk. (Of course, if the pitcher also knows these numbers, he might throw a more aggressive pitch instead.)

There are other indications that umpires don't want to insert themselves into the game. For as long as sports have existed, fans have accused officials of favoring star players, giving them the benefit of the calls they make. As it turns out, there is validity to the charges of a star system. Star players *are* treated differently by the officials, but not necessarily because officials want to coddle and protect the best (and most marketable) athletes. It happens because the officials don't want to influence the game.

If Albert Pujols, the St. Louis Cardinals' slugger—for our money, the best hitter in baseball today—is up to bat, an umpire calling him out on a third strike is likely to get an earful from the crowd. Fans want to see stars in action; they certainly don't want the officials to determine a star's influence on the game. Almost by definition, stars have an outsized impact on the game, so umpires are more reluctant to make decisions against them than, say, against unknown rookies. Sure enough, we find that on two-strike counts, star hitters—identified by their all-star status, career hitting statistics, awards, and career and current salaries—are much less likely to get a called third strike than are nonstar hitters for the exact same pitch location. This is consistent with omission bias and also with simple star favoritism.

But here's where our findings get really interesting. On three-ball counts, star hitters are *less* likely to get a called ball, controlling again for pitch location. In other words, umpires—already reluctant to walk players—are even more reluctant to walk star hitters. This is the opposite of what you would expect if umps were simply favoring star athletes, but it is consistent with trying *not* to influence the game. The result of both effects is that umpires prolong the at-bats of star hitters—they are more reluctant to call a third strike but also more reluctant to call the fourth ball. In effect, the strike zone for star hitters shrinks when they have two strikes on them but expands when they have three balls in the count. Umpires want star hitters in particular to determine their own fate and as a result give them more chances to swing at the ball.

As fans, we want that, too. Even if you root for the St. Louis Cardinals, you'd probably rather see Pujols hit the ball than walk. As an opposing fan, you'd like him to strike out, but isn't it sweeter when he swings and misses than when he takes a called third strike that might be ambiguous? We essentially want the umpire taken out of the play. Fans convey a clear message—*Let Pujols and the other team's ace duel it out*—and umpires appear to be obliging.

The umpire's omission bias affects star pitchers in a similar way. Aces are given slightly bigger strike zones, particularly on three-ball counts, consistent with a reluctance to influence the game by prolonging an outing. The more walks a pitcher throws, the more likely he is to be replaced, and that obviously has a sizable impact on the game and the fans.

■ ■ ■

In the NBA, home to many referee conspiracy theories, skeptical fans (and Dallas Mavericks owner Mark Cuban) have long asserted the existence of a "star system." The contention is that there is one set of rules for LeBron James, Kobe Bryant, and their ilk and a separate set for players on the order of Chris Duhon, Martell Webster, and Malik Allen. But confirming that star players receive deferential treatment from the refs is difficult, at least empirically. Stars have the ball more often, especially in a tight game as time winds down, and so looking at the number of fouls or turnovers on star versus nonstar athletes isn't a fair comparison. Unlike in baseball, where we have the Pitch f/x data, we can't actually tell whether a foul or violation *should* have been called. Did Michael Jordan push off against Bryon Russell before hitting the game-winning shot in the 1998 NBA finals? That's a judgment call, not a call that current technology can answer precisely and decisively.

The closest thing to a fair comparison between stars and nonstars we've found is what happens when two players go after a loose ball. A loose ball is a ball that is in play but is not in the possession of either team (think of a ball rolling along the floor or one high in the air). Typically, there is a mad scramble between two (or more) opposing players that often results in the referee calling a foul. We examined all loose ball situations involving a star and a nonstar player and analyzed how likely it is that a foul will be called on either one.* A nonstar player will be assessed a loose ball foul about 57.4 percent of the time, a star player only 42.6 percent of the time. If the star player is in foul trouble—three or more fouls in the first half, four or more fouls in the second half—the likelihood that he will be assessed a loose ball foul drops further, to 26.9 percent versus 73.1 percent for the nonstar. But what if the nonstar player is in foul trouble but the star isn't? It evens out, tilting slightly against the star player, who receives a foul 50.5 percent of the time, whereas his foul-ridden counterpart receives a foul 49.5 percent of the time. These results are consistent with the omission bias and the officials' reluctance to affect the outcome. Fouling out a player has a big impact on the game, and fouling out a star has an even bigger impact. Much like the called balls and strikes in MLB for star players, it is omission bias, not star favoritism, that drives this trend. Star players aren't necessarily being given better calls, just calls that keep them in the game longer.

---

* We define a "star" as any player in the top ten for receiving votes for MVP in any year, covering about 20 players. Star players for the years we examined were: Kobe Bryant, LeBron James, Allen Iverson, Shaquille O'Neal, Jason Kidd, Carmelo Anthony, Dwyane Wade, Vince Carter, Tim Duncan, Kevin Garnett, Yao Ming, Steve Nash, Dirk Nowitzki, Dwight Howard, Elton Brand, Tracy McGrady, Chris Paul, Amar'e Stoudemire, Kevin Durant, and Paul Pierce.

MAKE-UP CALLS

Another long-standing fan accusation against referees is the use of the make-up call. When an obviously bad call is made, the thinking goes, the officials soon compensate by making an equally bad call that favors the other team. Or, in the next ambiguous situation, the refs will side with the team that was wronged previously. A few years ago there was a commercial for Subway that featured a football ref standing at midfield and saying: "I totally blew that call. In fact, it wasn't even close. But don't worry. I'll penalize the other team—for no good reason—in the second half. To even things up."

■ ■ ■

The stats do seem to confirm the reality of make-up calls, but again, this stems from officials not wanting to inject themselves into the game. If you know you've made a bad call that influenced the game, you may be inclined to make a bad call in the other direction to balance it out. The hope is that "two wrongs make it right," but of course this means referees are consciously not always calling things by the rule book.

In baseball, we can look at make-up calls by the home plate umpire. If the umpire misses a strike call, how likely is it that the next pitch will be called a strike? It turns out that if the previous pitch was a strike but the umpire missed it and erroneously called a ball, the next pitch is much more likely to be called a strike even if it is out of the strike zone. If the previous pitch should have been called a ball but was mistakenly called a strike, the umpire is much more likely to call a ball on the next pitch even if the ball is in the strike zone. When umpires miss a called strike, they tend to expand their strike zone on the next pitch, and when they miss a called ball, they tend to shrink the strike zone on the next pitch.

The following graph shows the difference between the strike zones for pitches *immediately following* errant strike calls and errant ball calls. After an errant ball call, the strike zone magically

ACTUAL STRIKE ZONE
AFTER ERRANT STRIKE AND BALL CALLS

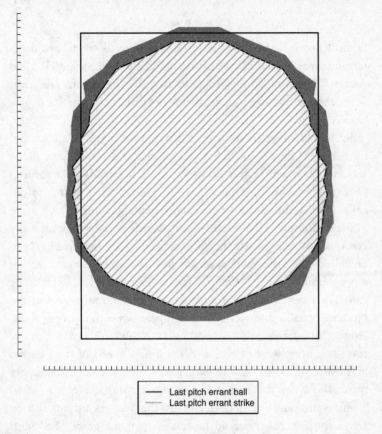

——— Last pitch errant ball
- - - Last pitch errant strike

grows by 70 square inches. This pattern holds even for the first two pitches of the at-bat.

Also, the more obvious the mistake, the more umpires try to make up for it on the next pitch. If the pitch was dead center down the plate and the ump failed to call a strike, he or she *really* expands the strike zone on the next throw. If the ball is way outside and the ump doesn't call a ball, he or she *really* tightens the strike zone the next time. Again, this is consistent with trying not to affect the game. Umpires are trying to balance out any mistakes they make, and the more obvious those mistakes are, the more they try to balance things out.

■ ■ ■

It's not just in MLB and the NBA that officials try to avoid determining the outcome. It also occurs in the NFL, the NHL, and soccer. The omission bias suggests that the rate of officials' calls will decrease as the game nears its conclusion and the score gets closer.

In the NBA there is some evidence that fouls are called less frequently near the end of tight games, especially in overtime. (That includes the intentional foul fest that usually attends close games.) However, by looking deeper into the *types* of fouls called, or not called, late in the game, we get a more striking picture. Fouls more at the discretion of the referee—such as offensive fouls, which any NBA ref will tell you are the hardest to call—are the least likely to be called when the game is on the line. For some perspective, on a per-minute basis, an offensive foul is 40 percent less likely to be called in overtime than during any other part of the game. Certain "judgment call" turnovers, too, disappear when the game is tight. Double dribbling, palming, and every NBA fan's favorite gripe, traveling, are all called half as often near the end of tight games and overtime as they are in earlier parts of the game. Remember the credo: When the game steps up, the refs step down.

But is this omission bias, or is it just that players are committing fewer fouls, turnovers, and mistakes when the game gets tight, and so referees have fewer calls to make? If we look at calls for which officials don't have much discretion, such as lost balls out of bounds (they have to call *something*), kicked balls, and shot clock violations, they occur at the same rate in the fourth quarter and overtime as they do throughout the game. In other words, players seem to be playing no more conservatively when the game is close and near the end.

One of our favorite examples of ref omission bias occurred in the championship game of the 1993 NCAA tournament, when Michigan's renowned Fab Five team played North Carolina. With

18 seconds to play and North Carolina leading by two points, Michigan star Chris Webber grabbed a defensive rebound and took three loping steps without dribbling. It was the kind of flagrant traveling violation that would have been cited in a church league game, but a referee standing just a few feet from Webber . . . did nothing. It was a classic case of swallowing the whistle. A traveling call would have doused the drama in the game. By overlooking Webber's transgression and declining to make a subjective call, the ref enabled the game to build to a dramatic climax. The no-call enraged Dean Smith, Carolina's venerable coach, who stormed down the court in protest. Billy Packer, the CBS commentator, was also apoplectic. "Oh, he walked!" Packer screamed. "[Webber] walked and the referee missed it!"

You might recall what happened next. Webber dribbled the length of the court. Then, inexplicably, he stopped dribbling and called time-out. Alas, Michigan had no time-outs left. Unlike a traveling violation, when a player motions for a time-out and his team has exhausted its ration, well, that's not a judgment call. That's a call an official *has* to make even in the waning seconds of an exhilarating championship game. And the officials did: technical foul. North Carolina wins.

In the NFL, more subjective calls (holding, illegal blocks, illegal contact, and unnecessary roughness) fall precipitously as the game nears the end and the score is close. But more objective calls (delay of game or illegal formation, motion, and shifts) are called at the same rate regardless of what the clock or scoreboard shows. The same is true in the NHL. More subjective calls (boarding, cross-checking, holding, hooking, interference) are called far less frequently at the end of tight games, but objective calls (delay of game, too many men on the ice) occur with similar frequency regardless of the game situation. We also find that in the NHL penalty minutes per penalty are lower late in the game. Referees have discretion over whether to call a major or a minor penalty—which dictates the number of minutes a player has to remain in the penalty box—and they are more reluctant to dispense more penalty minutes at the end of a tight game.

A European colleague snickered to us, "You wouldn't see this in soccer." But we did. We looked at 15 years of matches in the English Premier, the Spanish La Liga, and the Italian Serie A leagues. European officials are no better at overcoming omission bias than their American counterparts. Fouls, offsides, and free kicks diminish significantly as close matches draw to a close.

■ ■ ■

But refs aren't entirely to blame. As fans, we've come to expect a certain degree of omission bias, so much so that even the *right* call can be what the rules would suggest is the wrong call. Walt Coleman is the sixth-generation owner of Arkansas's Coleman Dairy, the largest dairy west of the Mississippi River. He is also an NFL official. (We told you these guys were exceptional.) Late in a 2002 playoff game between the Patriots and the Raiders, New England quarterback Tom Brady was sacked and appeared to fumble. After reviewing the play, Coleman, as referee, overturned the call and declared the pass incomplete, invoking the obscure "tuck rule" (NFL Rule 3, Section 21, Article 2, Note 2), which states:

> When [an offensive] player is holding the ball to pass it forward, any intentional forward movement of his arm starts a forward pass, even if the player loses possession of the ball as he is attempting to tuck it back toward his body. Also, if the player has tucked the ball into his body and then loses possession, it is a fumble.

The Patriots retained possession, scored a field goal on the final play of regulation, and won in overtime. Technically, Coleman appears to have made the correct call, but to many fans it didn't feel right to have an official insinuating himself into the game and going deep into an obscure part of the rule book at such a critical time. A decade later, the "tuck rule game" persists as one of the most controversial moments in NFL history. The "Tyree Catch," on the other hand, is hardly famous for its controversy. And the NFL's reaction was telling, too. The league did not offer Coleman up for a media tour the way they did Mike Carey.

For an even more vivid illustration of how fans and athletes expect officials to remove themselves during the key moments of sports contests, consider what happened at the 2009 U.S. Open tennis tournament. In the women's semifinal, Serena Williams, the 2008 defending champion, faced Kim Clijsters, a former top-ranked player from Belgium who'd retired from tennis to get married and start a family but had recently returned to make a spirited comeback. Although the draw sheet indicated that this was a semifinal match, the fans knew that it was the de facto final, pitting the two best players left in the tournament against each other. That Clijsters had beaten Serena's sister, Venus, a few rounds earlier infused the match with an additional layer of drama.

This was the rare sporting event that lived up to the considerable buildup. Points were hard fought. Momentum swung back and forth. As powerful as she was accurate, Clijsters won the first set 6–4. At 5–6 in the second set, Williams was serving to stay in the match. It was, as the cliché-prone might say, "crunch time." Clijsters won the first point. Williams won the next. Then Clijsters won a point to go up 15–30.

Two points from defeat, Williams rocked back and belted a first serve that landed a foot or so wide of the service box. The nervous crowd sighed. Williams bounced the ball in frustration and prepared to serve. After she struck her second serve but before the ball landed, the voice of a compactly built Japanese lineswoman, Shino Tsurubuchi, pierced the air: *"Foot fault!"*

Come again? A foot fault is a fairly obscure tennis rule dictating that no part of the server's foot touch—or trespass—the baseline before the ball is struck. (Imagine a basketball player stepping on the baseline while inbounding the ball.) Players can go weeks or even months without being cited for a foot fault violation. In this case, the violation was hardly blatant, but replays would confirm that it was legitimately a foot fault.

Williams lost the point as a result. The score was now 15–40, with Clijsters only a point from winning the game—and the match. As the crowd groaned, Williams paused to collect herself. Or so it seemed. Instead, she stalked over to Tsurubuchi, who was seated to the side of the court in, ironically, a director's chair. Then, in a ten-second monologue, Serena splintered whatever remained of tennis's facade as a prissy, genteel country club pursuit. Glowering and raising her racket with one hand and pointing a finger with the other, Serena barked: "You better be f—ing right! You don't f—ing know me! . . . If I could, I would take this f—ing ball and shove it down your f—ing throat!"

Having already been assessed a penalty for smashing her racket earlier in the match, Williams was docked a point. Since the foot fault had made the score 15–40, with the docked point the game and match were over. Bedlam ensued. Confused fans, shocked by the sudden end to the match, jeered and booed. Williams marched to the net, where officials were summiting, and protested. Slamming her racket, she walked over to Clijsters's side of the net, shook hands with her opponent, and then left the court. The blogosphere exploded. The "terrible tennis tirade" became a lead segment on CNN and front-page news internationally, the defining moment of the entire tournament.

Part of what made the episode so memorable was the kind of outrageous tirade one associates less with tennis than with, say, cage fighting. But it was also jarring to see an official essentially decide what had been a close, hard-fought contest between two worthy competitors. And in many corners, fans' outrage was directed at the official. How could the match be decided this way? We've come to expect omission bias in close contests. *Swallow the whistle!*

But wait, you say; the official didn't determine the outcome. Serena Williams did by her tirade, violating the rules. The lineswoman was simply doing her job. And if she had turned a blind eye to the violation, wouldn't she have been robbing Clijsters? Try telling that to John McEnroe. Commentating from the CBS broadcast booth that night, he remarked immediately: "You can't call that there! Not at that point in the match." One former NBA ref had the same reaction as he watched from his home. "Great feel for the match," he sarcastically texted a friend. Bruce Jenkins, a fine columnist for the *San Francisco Chronicle*, wrote, "[Tsurubuchi] managed to ruin the tournament . . . any sports fan knows you don't call a ticky-tack violation when everything is on the line."

A few weeks after Serena's Vesuvian eruption, *Sports Illustrated* readers voted her Female Athlete of the Decade, suggesting that the episode had done little to hurt her image. Tsurubuchi was less fortunate. She was hurriedly escorted from the stadium and flown back to Japan the next day. When we first attempted to interview her, we were told she was off-limits to the media. In fact, tennis officials wouldn't even disclose her name or confirm it when we learned it from other sources. (Compare this to the treatment Mike Carey received from the NFL after Super Bowl XLII.) Never mind that she made the correct call and didn't give in to omission bias. In effect, she was shamed for being right.

A full five months later, we finally caught up with Tsurubuchi at a small men's tennis event in Delray Beach, Florida, where she was working in anonymity. She cut a dignified, reserved figure, disappointed to have been recognized but too polite to decline a request to talk. Conversing with this reticent, petite woman—she looks to be about four foot eight—it was hard not to think of what calamity might have ensued if Serena Williams actually had acted on her threat that night. Her voice quivering as if on a vibrate setting as she recalled the incident that brought her unwanted fame, Tsurubuchi claimed that she'd had no choice. "I wish—I pray—for players: 'Please don't touch that line!'" she explained in halting English. "But if players [do], we have to make the call."

Would she make the same call again? "Yes," she said, looking dumbfounded. "It's tough and the players might not be happy . . . but the rules are the rules, no matter what."

Her call—her resistance to the omission bias to which we've

become accustomed in sports and in life—may have earned her widespread ridicule and disapproval, but she also won fans that night, including Mike Carey: "Making the hard call or the unpopular call, that's where guts are tested, that's the mark of a true official," he says. "You might have a longer career as an official if you back off. But you won't have a more accurate career."

GO FOR IT

Why coaches make decisions that reduce
their team's chances of winning

The sun retreated behind the hills on the west side of Little Rock
on a warm Thursday in September 2009. The Pulaski Academy
Bruins and the visiting Central Arkansas Christian Mustangs
emerged from their locker rooms and stretched out on the field
and applied eye black. Apple-cheeked cheerleaders alternated between practicing their routines and checking their backlog of text
messages. The air was thick with concession stand odor. The PA
blasted AC/DC's "Thunderstruck" and the predictable medley of
sports psych-up songs. A thousand or so fans found their seats on
the bleachers, filing past the placards for a store called Heavenly
Ham, Taziki's Greek Tavern, and other local businesses and insurance agents. It was conventional stuff, in other words, a typical
high school football tableau.

Then the game started.

On the first possession, Pulaski marched steadily downfield
until it faced fourth down and five at the Mustangs' 14-yard line.
The obvious strategy, of course, was to attempt an easy field goal
and be happy with a 3–0 lead. But without hesitation, the offense remained on the field and went for it. The quarterback, Wil
Nicks, rolled left, looked for a blue jersey, spotted one of his *five*

receivers, and zipped a swing pass near the sidelines that a junior receiver, Garrett Lamb, caught for a six-yard gain. First down.

A few plays later, thanks to an intentional grounding penalty and a bad snap, Pulaski faced fourth and goal from the opponent's 23-yard line. Again, conventional wisdom fairly screams: Attempt the field goal! Again, Pulaski did otherwise, going for it, lining up five receivers. Nicks was pressured out of the pocket and threw his *ninth* pass of the drive, a wayward throw, well behind the intended receiver, that fell innocuously to the turf. Central Arkansas Christian took over on downs.

By the end of the first quarter, the Bruins had declined to punt or attempt a field goal on all four of their fourth downs, field position be damned. Then again, this wasn't so surprising given that the team's roster listed neither a punter nor a kicker among its 45 players. Nicks, the quarterback, had already attempted 15 passes, on a pace to eclipse the 50 tosses he'd thrown in his previous game.

Early in the second quarter, Pulaski scored its first touchdown. After a nifty play fake, Nicks threw over the defense to a streaking receiver, Caleb Jones. On the ensuing kickoff, eleven Pulaski players massed near the 40-yard line. With the ball propped horizontally on the tee, resembling an egg on its side, the Pulaski players ran in different directions, as if performing an elaborate dance for which only they knew the choreography. With the play clock winding down, a burly senior tackle, Allen Wyatt, squirted a nine-yard kick that hugged the turf and bounced awkwardly before the visiting team pounced on the ball and hugged it like a long-lost relative.

As one of the texting cheerleaders might have abbreviated it: WTF? Who ever heard of deploying an onside kick in the second quarter, much less when you aren't behind?

But none of it provoked surprise among the Pulaski fans. After the opponents fell on the ball, the Bruins jogged off as if nothing remarkable had happened. And in retrospect, nothing had. Turns out that after most of Pulaski's touchdowns, the team went for a two-point conversion, not an extra point. On kickoffs, either they attempted fluttering onside kicks from any of a dozen formations or the designated kicker—who's not really a kicker—would turn sideways and purposely boot the ball out of bounds, preventing a return.

And the, um, avant-garde play-calling didn't stop there. When Central Arkansas Christian punted, Pulaski didn't position a man back, much less attempt a return. Instead, it chose to let the ball simply die on the turf. Pulaski threw the football on the majority of downs—except for third and long, when they often ran the ball. They sometimes lined up eight men on one side of the field. From a spread offense formation, they deployed crafty shuffle passes, direct snaps to the running back, end arounds, reverses, and an ingenious double pass. Pulaski often showed greater resemblance to a rugby team than to a football team.

The players, not surprisingly, love it. What teenager who goes out for the high school football team wouldn't be enthralled with a system that encourages passing on most downs, routinely racks up 500 yards a game in total offense, and is chock full of trick plays? "You can't imagine how fun it is," gushed Greyson Skokos, a thickly proportioned running back and one of four Bruins players who would go on to catch at least 50 passes in the 2009 season.

The defensive players don't mind it, either. Though they're not on the field much, they welcome the challenge that comes when the offense fails to convert a fourth down and the opponent suddenly takes possession of the ball in the "red zone," sometimes just a few yards from scoring. The Pulaski fans are accustomed to it by now, as well. Most enjoy the show, shake their heads, and almost uniformly refer to the team's coach, Kevin Kelley, as a "mad scientist."

Truth is, Kelley isn't mad at all. Quite the opposite. He's relentlessly rational, basing his football philosophy not on whimsical experimentation or hot spur-of-the-moment passion but on cool thinking and cold, hard math.

Playing high school ball in Hot Springs, Arkansas, in the 1980s, Kelley watched in frustration as his conservative coach ordered the team to run on first and second downs, pass on third down, and punt or attempt a field goal on fourth down. To Kelley it made no sense: "It was like someone said, 'Hey, it's fourth down, you have to punt now.' So everyone started doing it without asking why. To me, it was like, 'You can have an extra down if you want it. No, I'll be nice and just use three.'" At college at Henderson State, Kelley took a few economics courses, and though demand and supply curves didn't captivate him—he ended up majoring in PE—he was intrigued by the thought of applying basic statistics and principles of economics to football. Within a few years, he had his chance. In 2003, he was promoted to head football coach at Pulaski Academy, an exclusive private school where Little Rock's prominent families sent their kids. He decided to amass statistics and, based on the results, put his math into practice.

Among his early findings: His teams averaged more than six yards per play. "Think about it," he says. "[At six yards per play] if you give yourself four downs, you only need two and a half yards per down. You're in great shape. Even if you're in, like, third and eight, you should be okay. I'll keep all four downs, thank you very much!" Kelley also realized quickly that using all four downs and breaking with hidebound football "wisdom" confused defenses, enabling his team to gain even more yards. "When third and seven is a running down and fourth and one could be a passing down, and defenses don't know whether to use dime packages or nickel packages, the offense does even better."

Although Pulaski is hardly successful on every fourth-down attempt, it succeeds roughly half the time, enough to convince Kelley that statistically, his team is better off going for it every time. And keep in mind that this is *without* the element of surprise.

According to Kelley's figures, in Arkansas high school football, teams tend to average a touchdown on one of every three possessions. By punting away the ball three times when he didn't have to, he'd essentially be giving the opponents a touchdown each game.

By the time Pulaski played Central Arkansas Christian in September 2009, it had been more than two years since one of his teams had attempted a punt—and that was a gesture of sportsmanship to prevent running up the score. (Still more proof that no good deed goes unpunished, it was returned for a touchdown, cementing Kelley's belief that punting is a flawed strategy.) Again, Kelley and his numbers: "The average punt in high school nets you around 30 yards, but especially when you convert around half your fourth downs, it doesn't make sense to give up the ball," he says. "Honestly, I don't believe in punting and really can't ever see doing it again."

He means *ever*. What about the most extreme scenario, say, when the offense is consigned to fourth and long, pinned near its own end zone? It's still better not to punt? "Yup," he says, arms folded across his thick belly. Huh?

According to Kelley's statistics, when a team punts from that deep, the opponent will take possession inside the 40-yard line and, from such a favorable distance, will score a touchdown 77 percent of the time. Meanwhile, if the fourth-down attempt is unsuccessful and the opponent recovers on downs inside the 10-yard line, it will score a touchdown 92 percent of the time. "So [forsaking] a punt you give your offense a chance to stay on the field. And if you miss, the odds of the other team scoring a touchdown only increase 15 percent."

The onside kicks? According to Kelley's figures, after a conventional kickoff, the receiving team, on average, takes over at its own 33-yard line. After an unsuccessful onside kick, it assumes possession at its own 48. Through the years, Pulaski has recovered between one-quarter and one-third of its onside kicks. "So you're giving up 15 yards for a one-in-three chance to get the ball back," says Kelley. "I'll take that every time!"

The decision not to return punts? In high school, punts seldom travel more than 30 yards. And at least for a small, private high school where speed demons are in short supply, Pulaski's return team seldom runs back punts for touchdowns. A far more likely outcome for the return team is a penalty or a fumble. So Kelley—the same man who will go for it on fourth and 20— instructs his team to avoid returning punts altogether. "It's just not worth the risk," he explains.

A folksy, exceedingly likable man in his mid-forties whose wife, kids, and elderly mom come to every Pulaski home game, Kelley makes no pretenses about his academic credentials. "I just like to quantify it all together," he says. "But I'm not like an astrophysicist or a real math whiz."

The real math whizzes, however, confirm much of Kelley's analysis. David Romer, a prominent Cal–Berkeley economist and member of the National Bureau of Economic Research—whose wife, Christina, chaired President Obama's Council of Economic Advisers for two years—published a 2005 study titled "Do Firms Maximize? Evidence from Pro Football." Taking data from the first quarter of NFL games, Romer concluded that in many fourth-down situations, statistically, teams are far better off forgoing a punt or field goal and keeping the offense on the field for another down. His paper is filled with the kind of jargon that would induce narcolepsy among most football fans. He also looked only at first-quarter results because he figured his data would be skewed by obvious fourth-down attempts, for example, when a team is down by seven points late in the game and everyone knows it has to go for it. But, greatly simplified, here are his conclusions:

- *Inside the opponent's 45-yard line, facing anything less than fourth and eight, teams are better off going for it than punting.*
- *Inside the opponent's 33-yard line, they are better off going for it on anything less than fourth and 11.**
- *Regardless of field position, on anything less than fourth and five, teams are* always *better off going for it.*

Other mathematicians and game theory experts have reached similar conclusions. Frank Frigo and Chuck Bower—a former backgammon world champion and an Indiana University astrophysicist—created a computer modeling program for football called ZEUS that takes any football situation and furnishes the statistically optimal strategy. The results often suggest going for it when the conventional football wisdom says to punt.

---

* The exception: if little time remains and a field goal would decide the game.

Kelley believes that the "quant jocks" don't go far enough to validate the no-punting worldview and, more generally, the virtues of risk-taking. "The math guys, the astrophysicist guys, they just do the raw numbers and they don't figure emotion into it—and that's the biggest thing of all," he says. "The built-in emotion involved in football is unbelievable, and that's where the benefits really pay off." What he means is this: A defense that stops an opponent on third down is usually ecstatic. They've done their job. The punting unit comes on, and the offense takes over. When that defense instead gives up a fourth-down conversion, it has a hugely deflating effect. At Pulaski's games, you can see the shoulders of the opposing defensive players slump and their eyes look down when they fail to stop the Bruins on fourth down.

Conversely, Kelley is convinced that fourth-down success has a galvanizing effect on the offense. "It was do or die and they did," he says. "I don't think it's a coincidence that on more than half of our touchdown drives, we converted a fourth down."

Similarly, according to Kelley's statistics, when an Arkansas high school team recovers a turnover, it is almost twice as likely to score a touchdown as it is when it receives a punt at the same yard line. He cites this as another argument in support of onside kicking and the refusal to risk fumbling a punt return.

The benefits of Kelley's unique system don't stop there. Because the formations and play-calling are so out of the ordinary, Pulaski tends to induce an inordinate number of penalties from the opposing team. Since Pulaski's ways are so thoroughly unique, in the week before playing the Bruins, opponents depart from their normal preparation routine. They devote hours to practicing all manner of onside kick returns and defending trick plays and installing dime packages on fourth down. There's that much less time to spend practicing their own plays.

Especially in high school, when off-season practice time is limited—and you're dealing with teenage attention spans—those lost hours can be critical. In the run-up to the Pulaski game, Central Arkansas Christian's coach, Tommy Shoemaker, estimated that he spent half his practices worrying about the Bruins' schemes. How much time did his team usually spend on the opposition? "Maybe twenty percent." Then again, he added wryly, at least his boys didn't have to spend time worrying about punt returns or field goal blocks. Turning serious, he added: "Keep in mind, we play these guys every year. I couldn't imagine what it'd be like getting ready if you didn't have any history."

Still another abstract benefit of playing for Pulaski: The experience is so different from traditional high school football that the Bruins' players feel as though they're part of something unique, an elite unit amid regular cadets. The team bonds have solidified; the offensive and defensive players consider themselves kindred spirits, bracketed together by their singular coach. And there are so many trick plays and intricate formations that players, by necessity, are alert at all times.

Happy as Kelley is to unleash his empirical evidence, these are the numbers that matter most to him: In the years since he took over as head coach, Pulaski is 77–17–1 through 2009, winning 82 percent of its games, and has been to the state championship three times, winning twice. All this despite drawing talent from only a small pool of private school adolescents. "I'm telling you," says Kelley. "It works."

■ ■ ■

It's up for debate whether Kelley's operating principles would work in all cases, for all teams, on all levels—for the record, he thinks they would—but his success at Pulaski is beyond dispute. With that record, you'd think other coaches would try to implement some form of Kelley-ball, but although he has become a minor celebrity in coaching circles and speaks at various banquets and conferences, he has not been flattered sincerely by imitation. Other coaches have cribbed the West Coast offense from Bill Walsh, the former Stanford and San Francisco 49ers coach, or the spread formation from Mike Leach, late of Texas Tech, but Kelley draws little more than a curious eye. "If there's another team out there that don't ever punt," he says with a shrug, "I haven't heard of 'em."

Several years ago, a prominent college coach paid a visit to Kelley's office at Pulaski, a nondescript box off to the side of the basketball court. The coach—Kelley doesn't want to name him for fear it might hurt the future recruitment of Pulaski players—asked for a primer on "that no punting stuff." Kelley happily obliged, explaining his philosophy and showing off his charts. "He wrote all sorts of stuff down in this big old binder and I'm thinking, 'Finally someone else sees the light.'" But when Kelley watched the coach's team play the next season, he saw no evidence that he had a disciple. Even armed with the knowledge that he was disadvantaging his team by his decision to punt, the coach routinely ordered the ball booted on fourth down.

That mirrors David Romer's experience. In his paper, Romer, the Berkeley economist, argued that the play-calling of NFL teams shows "systematic and clear cut" departures from the decisions that would maximize their chances of winning. Based on data from more than 700 NFL games, Romer identified 1,068 fourth-down situations in which, statistically speaking, the right call would have been to go for it. The NFL teams punted 959 times. In other words, nearly 90 percent of the time, NFL coaches made the suboptimal choice.

Inasmuch as an academic paper can become a cult hit, Romer's made the rounds in NFL executive offices, but most NFL coaches seemed to dismiss his findings as the handiwork of an egghead, polluting art with science. Plenty admit to being familiar with Romer's work; few have put his discoveries into practice.

It all lays bare an abiding irony of football. Here are these modern-day gladiators, big, strong Leviathans. It's a brutal, unforgiving game filled with testosterone and bravado. Players collide off each other so violently that there might as well be those cartoon bubbles "Pow" and "Bam." The NFL touts itself as the baddest league of all. Yet when it comes to decision-making, it's remarkably, well, wimpy.

There's not just an aversion to risk and confrontation; coaches often make the *wrong* choice. In other words, they're just like . . . the rest of us.

Time and again, we let the fear of loss overpower rational decision-making and often make ourselves worse off just to avoid a potential loss. Psychologists call this loss aversion, and it means we often tend to prefer avoiding losses at the expense of acquiring gains. The psychologists Daniel Kahneman and Amos Tversky are credited with discovering this phenomenon. (Kahneman won the Nobel Prize for this work in 2002; Tversky died in 1996 before being recognized.) As the late baseball manager Sparky Anderson put it: "Losing hurts twice as bad as winning feels good."

For most of us, the pain of losing a dollar is far more powerful than the pleasure of winning a dollar. In a frequently cited psychology experiment, subjects are offered two gambles that have identical payoffs but are framed differently. In the first gamble, a coin is flipped, and if it lands heads, you get $100; if tails, you get nothing. In the second gamble you are given $100 first and then flip the coin. If the coin lands heads, you owe nothing; if tails, you pay back the $100. Subjects dislike the second experiment much more than the first even though the actual gains and losses are identical.*

Marketing and advertising execs cater to this bias. Would you rather get a $5 discount or avoid a $5 surcharge? The same change in price framed differently has a significant effect on consumer behavior. A study of insurance policies, for instance, found that consumers switch companies twice as often when their carrier raises rates, as opposed to when the competition decreases its rates by the same amount. In everyday life, loss aversion causes people to make suboptimal choices. Many home owners looking to sell their houses right now would rather keep them on the market for an extra year than drop the price to $5,000 less than they paid, even though keeping the home for an extra year will surely cost them more than

---

* Research even shows that the brain processes losses differently from gains. In experiments offering individuals different gambles with the same payoff, but with one framed in terms of gains and the other in terms of losses, researchers at UCLA—Sabrina M. Tom, Craig R. Fox, Christopher Trepel, and Russell Poldrack—found that a number of areas in the brain showed increasing activity as potential gains increased, whereas potential losses showed decreasing activity in these same areas, even though the actual dollars won and lost were the same.

$5,000. A study of home sales by two economics professors, David Genesove and Christopher Mayer, then at the University of Pennsylvania's Wharton School of Business, showed this pattern. Home owners were reluctant to reduce the sale price below what they paid for the house even when continuing to own it meant incurring carrying costs—mortgage, utilities, maintenance—far exceeding the reduction in price needed to sell it. The idea of a loss was just too painful for them. In contrast, home owners facing a gain on a house often sold too early and for too little. The gain didn't matter as much as long as there wasn't a loss.

On Wall Street, fear of loss is often behind dubious investment strategies. Mutual fund managers, for example, will hold well-known or recognizable companies instead of obscure companies that are expected to deliver much better performance. The rationale: If you lose money by buying Walmart or Microsoft—recognizable blue chip companies—no one will blame you. You won't get fired; they'll chalk it up to "bad luck." Even though a small, obscure company might be a better bet, on the off chance that it doesn't pay off, you risk losing the client. So it is that many mutual fund managers will choose good companies over good investments.

On the television reality show *The Biggest Loser,* obese contestants compete to lose weight. The more they lose, the more they are rewarded. Two Yale professors, Ian Ayres, an expert in contract law, and Dean Karlan, a behavioral economist, were desperate to lose weight. Like the *Biggest Loser* contestants, they tried to find motivation in rewards. It didn't work, and so they flipped the *Biggest Loser* concept around and tried to motivate themselves with loss aversion. They entered a weight-loss bet with each other, and each one committed to pay the other $1,000 a week if he didn't lose the required weight. In addition, once the weight was lost, it couldn't be gained back without incurring the $1,000 penalty.

Two years later, neither professor has seen a dime of the other's money—and they've lost almost 80 pounds between them. They launched a company, stickK.com, to help people facilitate personal commitment contracts for weight loss and other personal goals by using loss aversion. If you don't live up to your end of the

contract, they give your money to charity or a designated benefi-
ciary. (In another variation, the losers have to donate the money
to a cause that runs counter to their political sensibilities: gun hat-
ers contributing to the NRA, pro-lifers contributing to Planned
Parenthood.)

This same loss aversion affects coaches. They behave much like
the shortsighted mutual fund manager who forgoes long-term
gains to avoid short-term losses and the amply girthed professors
who could lose weight only when faced with a loss rather than a
reward. The coaches are motivated less by potential gain (a touchdown) than by fear of a concrete loss (the relative certainty of
points from a field goal).

More broadly, many coaches ultimately are motivated less by the
potential of a Super Bowl ring than by the potential loss of something valuable they possess: their job. And in sports, there are few
faster ways to lose your employment than by bucking conventional
thinking, by trying something radical, and failing. A coach order-
ing his team to punt on fourth and three—even when it's statistically
inadvisable—faces little backlash. He is the money manager who
plays it safe and loses with Walmart. If he goes for it and is unsuccessful, there's hell to pay. He is then the money manager who loses
on that unknown tech stock and now risks losing the entire account.

It makes for an odd dynamic in which the incentives and objectives of coaches aren't perfectly aligned with those of the owners
or the fans. All want to win, but since the owners and fans can't
be fired, they want to win at all cost. Give a coach truth serum and
then ask what he'd prefer: go 8–8 and keep your job or go 9–7
and, because of what's perceived to be your reckless, unconventional play-calling, lose your job?

■ ■ ■

It's not just football coaches who make the wrong choices rather
than appear extreme. In basketball, for instance, prevailing wisdom dictates that coaches remove a player with five fouls, particularly a star, rather than risk having him foul out of the game. But
does this make sense?

We can start by measuring how long a player sits on the bench once he receives a fifth foul. We analyzed almost 5,000 NBA games from the 2006–2007 to 2009–2010 seasons and found that when a player receives his fifth foul, on average, there is 4:11 left to play in the game. He's benched for about 3:05 of that remaining time, leaving only 1:06 of actual playing time with five fouls. Stars are treated a little differently.* On average, they don't receive their fifth foul until there is 3:44 left, and coaches bench them for a little more than two minutes.

The strategy of sitting a player down with five fouls and waiting until the end of the game to put him back in presumes that players, particularly stars, are more valuable at the end of the game than at other times. But this is seldom the case.

Statistical analysts in basketball have created "plus-minus," or an "adjusted plus-minus," a metric for determining a player's worth when he is on the floor. Simply put, it measures what happens to the score when any particular player is on the court. When a player is plus five, that means his team scored five more points than the opponent when he was on the floor. Thus, this measure takes into account not only the individual's direct influence on the game from his own actions but also the indirect influence he has on his teammates and his opponents. It measures his net impact on the game.

As often as we hear about "clutch players," for the average NBA player, his contribution to the game, measured by plus-minus, is actually almost two points *lower* in the fourth quarter than in the first quarter. This is also true for star players and is even the case in the last five minutes of the game. Thus, the strategy of sitting a player down with five fouls to save him for the end of the game seems to be based on a faulty premise—he is no more valuable at the end of the game.

Now consider who replaces the player when he sits on the bench. The average substitute summoned in the fourth quarter to

* Stars are defined as players receiving votes for MVP that season or All-Star players.

replace the teammate in foul trouble, not surprisingly, has an even smaller impact. Replacing the star player in foul trouble with a sub has the net effect of reducing the team's points by about 0.17 for every minute the star is on the bench. This is a heavy price to pay. (We considered that a star player in foul trouble might compete conservatively, so maybe the difference between a sub and a star who plays conservatively with five fouls isn't all that great. But no, it turns out that's not true. If anything, star players have an even *higher* plus-minus than normal when they are in foul trouble.)

Leave a player with five fouls in the game and what happens? The average player with five fouls will pick up his sixth and foul out of the game only 21 percent of the time. A star is even less likely to pick up a sixth foul (only 16 percent of the time once he receives his fifth foul; remember "Whistle Swallowing"?). Thus, leaving a player in the game with five fouls hardly guarantees that he'll foul out.

Bottom line: An NBA coach is much better off leaving a star player with five fouls in a game. By our numbers, coaches are routinely giving up about 0.5 points per game by sitting a star player in foul trouble (and that doesn't include the minutes he might have sat on the bench with three fouls in the first half). That may not seem like much, but in a close game, in which these situations often occur, it could mean the difference between winning and losing. We estimate that leaving a player in with five fouls instead of benching him improves the chances of winning by about 12 percent. Over the course of a season, this can mean an extra couple of wins. Yes, a player may foul out of a game, but benching the player *ensures* that he's out of the game. As Jeff Van Gundy, former coach of the Houston Rockets and New York Knicks and current television announcer, once put it on the air, "I think *coaches* sometimes foul their players out."

So why don't NBA coaches let their players—particularly their stars—keep playing when they have a lot of fouls? Again, loss aversion and incentives. If you lose the game by following convention and sitting your player down, you escape the blame. But if you play him and he happens to foul out and the team loses, you guarantee yourself a heaping ration of grief on sports talk radio, in columns, and over the blogosphere even though the numbers strongly argue in favor of leaving the player in the game. As with punting on fourth down, coaches are willing to give up significant gains to mitigate the small chance of personal losses. Presented with this evidence, one NBA coach maintained that he was still going to remove a player when he picked up his fifth foul late in the game. Why? "Because," he said, "my kids go to school here!"

■ ■ ■

Another example of loss aversion is seen in baseball. Game after game, the same scene plays out with almost numbing familiarity: It's the ninth inning, the manager for the winning team summons the liveliest arm in the bullpen, the PA system cranks up ominous music—Metallica's "Enter Sandman" more often than not—and out trots Mariano Rivera, the Yankees' peerless relief pitcher, or his equivalent, to record the save. Why? Because conventional baseball wisdom dictates that managers use their best relief pitchers at the *end* of games to preserve victories. The presumption: This is the most important part of the game, with the greatest impact on the outcome. Not for nothing are these pitchers called closers.

But where is it written that a closer must close? What if the most important moment in the game, when the outcome is most likely to be affected, occurs earlier? Might it not make more sense to summon Rivera or Boston Red Sox closer Jonathan Papelbon when the game is tied in the sixth inning and there are runners on base? Wouldn't they be more valuable at this juncture than they are when they usually report to work: the ninth inning when their team is ahead?

Yet you almost never see a manager use his bullpen ace before the eighth inning. Why? Because, again, what manager wants to subject himself to the inevitable roasting if this strategy fails? If your closer isn't available to seal the game and you happen to lose . . . well, managers have been fired for lesser offenses. (Keep in mind, too, that closers like to accumulate "saves"—which occur if they are the last one pitching—since saves translate into dollars in the free agent market.)

Even in hockey, one can see loss aversion affecting coaching strategy. "Pulling the goalie" and putting another potential goal scorer on the ice near the end of a game when your team is losing decidedly improves your chances of scoring a goal and tying the game, but it also increases the risk that with the net empty, an opponent will score first and put the game out of reach. We found that NHL teams pull their goalies too late (on average with only 1:08 left in the game when down by one goal and with 1:30 left when down by two goals). By our calculations, pulling the goalie one minute or even two minutes earlier would increase the chances of tying the game from 11.6 percent to 17.6 percent. Over the course of a season that would mean almost an extra win per year. Why do teams wait so long to pull the goalie? Coaches are so averse to the potential loss of an empty-net goal—and the ridicule and potential job loss that accompany it—that they wait until the last possible moment, which actually reduces their chances of winning.

When *do* we see coaches take risks? Well, when do we take risks in everyday life? Usually when there's little or nothing to lose. You're less likely to be loss-averse when you *expect* to lose. Think of your buddy in Vegas who's getting crushed at the tables. Already down $1,000, he'll take uncharacteristic risks, doubling down when he might otherwise fold, in hopes of winning it back. How many times have you gotten lost driving the back roads and taken a few turns based on intuition rather than consult your map or GPS? "Hey, why not? I'm lost already." For that matter, how many schlubs have overreached around the time of last call, figuring that if they get shot down, they're no worse for it?

Coaches are subject to the same thinking: In the face of desperation, or a nearly certain loss, they'll adopt an unconventional strategy. They'll go for it on fourth down when their team is trailing late in the game. They'll pull the goalie with a minute left. They'll break the rotation and use their ace pitcher in the seventh game of a World Series. Why not?

Consider how the forward pass became a part of football. It was legalized in 1906 but hardly ever deployed until 1913, seven years later, when a small, obscure Midwestern school, Notre Dame, had to travel east to face mighty Army, a heavily favored powerhouse. With little to lose, the Fighting Irish coach, Jesse Harper, decided to employ this risky, newfangled strategy by using his quarterback, Charlie "Gus" Dorais, and his end, a kid named Knute Rockne. The summer before, Dorais and Rockne had been lifeguards on a Lake Erie beach near Sandusky, Ohio, who passed the time throwing a football back and forth. The Army players were stunned as the Irish threw for 243 yards, which was unheard of at the time. Notre Dame won easily, 35–13. After that, the Irish no longer resided in college football obscurity, Dorais and Rockne became one of the first and best passing tandems of all time, and the forward pass was here to stay. Dorais and Rockne would both go on to become revered Hall of Fame coaches, in large part because they continued deploying their passing tactics at the coaching level.

■ ■ ■

In the rare instances when coaches in sports embrace risk systematically—not in the face of desperation but as a rule—there is a common characteristic. It has nothing to do with birth order or brain type or level of education. Rather, those coaches are secure in their employment. If the experiment combusts, they have little to lose (i.e., their jobs).

Is it coincidence that New England Patriots coach Bill Belichick opts to go for it on fourth down more often than any of his colleagues do? True, Belichick is a cerebral sort who understands risk aversion and probability as well as anyone, but he's also won three Super Bowls since 2001 and has more job security than any other coach in the NFL. We noticed that before he became a coaching star, Belichick approached the game quite differently. In his first head coaching stint in the NFL, with Cleveland, Belichick amassed an unimpressive 45 percent winning percentage and had only one winning season in five years. In Cleveland, he never exhibited the penchant for risk-taking that he shows with the Patriots. Back when he commanded the Browns, he went for it on fourth down only about one out of seven times. Since taking the helm at New England in 2000, Belichick has gone for it on fourth down a little more than one in four times.

But this tells only part of the story. In Cleveland, Belichick's team trailed more often, and so many of the fourth downs he went for were in desperate situations—trailing near the end of the game. In New England, he had better teams and hence was ahead much more frequently, facing fewer "desperate" fourth-down situations. Looking only at fourth-down situations in the first three quarters with his team trailing by less than two touchdowns, we found that in Cleveland he went for it on fourth down only about one in nine times, but in New England he went for it about one out of four times in the same situations. Belichick was almost three times more likely to go for it on fourth down in New England than he was in Cleveland.

One could argue that having a better team in New England meant he was more likely to convert more fourth downs, which is why he chose to go for it more often. True, his Patriots converted more of their fourth-down attempts than his Browns did, but the differences weren't big (59 percent versus 51 percent), certainly not three times larger. Plus, in Cleveland, since he attempted more "desperate" fourth downs, sometimes with more than ten yards to go, you'd expect the success rate to be lower. Controlling for the *same* yardage, Belichick's Patriots were only slightly better than his Browns at succeeding on fourth down.

So what changed his appetite for risk? Belichick didn't have great job security in Cleveland, as evidenced by his eventual dismissal in 1996. Even in New England the first couple of years, when his job was less certain, he remained conservative. Only after his teams had won multiple Super Bowls and he was hailed as "the smartest coach in football" did his risk-taking increase. His job security at that point wasn't an issue.

But even a secure coach bucks convention at his own peril. In November 2009, the Indianapolis Colts, undefeated at the time, hosted the New England Patriots. The latest installment in the NFL's most textured rivalry, it was a Sunday night affair televised on NBC. New England led comfortably for most of the game, but in the fourth quarter the wires of the Colts' offense started to connect. Indianapolis scored a late touchdown to close the score to 34–28. The crowd noise at Lucas Oil Stadium reached earsplitting levels.

On the Patriots' next possession, they moved the ball with deliberate slowness and faced fourth and two on their own 28-yard line. It was a compelling test case for risk management in the NFL. If the Patriots punted, it was a virtual certainty that Indianapolis would get the ball back, leaving Peyton Manning slightly more than two minutes and two time-outs (one of their own and one from the two-minute warning) to move the ball 65 or 70 yards to score a touchdown—a feat he had achieved on many occasions, including the last time the two teams had met in Indianapolis.

If the Patriots went for it and converted, the game's outcome would effectively be sealed. However, if the Patriots went for it and failed, they would give the Colts the ball inside their 30-yard line. So going for it would either end the game or—worst-case scenario—give the ball back to Manning and the Colts' offense 35 to 40 yards closer than punting the ball would. If the Colts scored a touchdown quickly from that shorter distance, there might still be time for the Patriots to kick a game-winning field goal. There were other factors as well. The Patriots' defense was visibly exhausted, and, thanks to injuries, two starters were missing from New England's defensive secondary, another factor militating against punting. Watching the game at home in Arkansas, Kevin Kelley shouted at his television, hoping Belichick would have the "guts" (his word) to forsake punting.

Beyond gut intuition, the analytics also supported going for it. Crunching the numbers, the average NFL team converts on fourth and two about 60 percent of the time. If successful, the Patriots would almost assuredly win the game. If they failed and the Colts took over on the Pats' 30-yard line with two minutes left and down by six points, the Patriots were still 67 percent likely to win the game. In other words, the Colts had only a one in three chance of actually scoring a touchdown from the Patriots' 30, so it was hardly as if the Patriots were conceding a touchdown if the fourth-down attempt failed. Alternatively, punting the ball would put the Colts at roughly their own 30, which gave the Patriots about a 79 percent chance of winning. There was, then, only a 12 percent difference in the probability of winning the game if the Patriots failed on fourth down versus if they punted the ball. And if they converted (which was 60 percent likely), the game would effectively be over. Adding everything up, going for it gave the Patriots an 81 percent chance to win the game versus a 72 percent chance if they punted.* Even tweaking these numbers by using different assumptions, you'd be hard-pressed to favor punting. At best, you could say it was a close call between punting and going for it; at worst, going for it dominated.

NFL fans probably will recall what happened next. Belichick ordered his offense to stay on the field. "We thought we could win the game on that play," he said afterward. New England's quarterback, Tom Brady, had thrown for nearly 400 yards that evening but couldn't pick up the crucial 72 inches on fourth down. He zipped a quick pass to Kevin Faulk. Like a man smushing out a cigarette in an ashtray, Colts safety Melvin Bullitt ground Faulk into the turf a few feet shy of the line.

By then, the fates had already written the script. As condemnation of Belichick's "cowboy tactic" and "needless gamble" was beginning to crackle in the broadcast booth and on the blogosphere, the Colts marched methodically, inevitably, to the end

---

* These numbers are based on league averages for the probability of scoring a touchdown from a specific field position and the probability of converting a fourth and two. It turns out the Patriots are much more likely than the average team to convert fourth and two (70 percent versus 60 percent) and the Colts, with Peyton Manning, are much more likely to score a touchdown than the average team from most positions on the field. But these two effects probably cancel each other out. One other thing to consider, however, that would also favor going for it over punting is the fact that the Patriots probably would adopt a more conservative defensive strategy or "prevent" defense to guard against the deep ball if the Colts started on their own end of the field. This probably would allow Peyton Manning to march quickly down to the Patriots' end of the field in less time than usual, making the decision to punt even less valuable.

zone. With seconds to play, Indianapolis scored a touchdown on a one-yard pass to win the game 35–34.

Belichick may have been the most highly regarded coach in the NFL and may have made what was, statistically anyway, the correct call, but out came the knives. The reviews from the salon were brutal:

- *"You have to punt the ball in that situation. As much as you may respect Peyton Manning and his ability, as much as you may doubt your defense, you have to play the percentages and punt the ball. . . . You have got to play the percentages and punt the ball."* –NBC analyst Tony Dungy, the Colts' former coach
- *"It was a really bad coaching decision by Coach Belichick. I have all the respect in the world for him, but he has to punt the ball. The message you send in the locker room is, 'I have no confidence in my young guys on defense.'"*–former Patriots safety and current NBC analyst Rodney Harrison
- *"Ghastly. . . . Too smart for his own good this time. The sin of hubris."*–Boston Globe *columnist Dan Shaughnessy*
- *"Is there an insanity defense for football coaches?"*–Boston Herald *columnist Ron Borges*
- *"I hated the call. It smacked of 'I'm-smarter-than-they-are' hubris. This felt too cheap."*–Peter King, SI.com
- *"My vocabulary is not big enough to describe the insanity of this decision."*–former NFL quarterback and ESPN analyst Trent Dilfer
- *"Fourth-and-jackass. That's our name for a now-infamous play in New England Patriots' history."*–Pete Prisco, CBSSports.com
- *"So what was more satisfying Sunday night, watching good guy Peyton Manning rally the Colts or bad guy Bill Belichick choke as a tactician?"*–Jeff Gordon, St. Louis Post-Dispatch

Of course none of these criticisms mentioned that punting was statistically inferior or at best a close call relative to going for it. In fact, they claimed the opposite, that punting was the superior strategy. It wasn't.

■ ■ ■

It wasn't just that the Patriots had lost. It was that Belichick had dared to depart from the status quo. He was the geek with the pocket protector, and damn if it didn't feel good when he was too smart for his own good. It had all the ring of the cool kids in school celebrating when the know-it-all flunked the test.

Unless blessed with clairvoyance, you make a decision before you know the outcome. The decision to go for it was the *right* decision. That it didn't work out doesn't change the soundness of the decision. Yet people seldom see it this way. They have what psychologists call hindsight bias. If you did the right thing but failed because of bad luck, you're stupid. If you did the wrong thing but succeeded because of good fortune, you're a genius. Of course, it's often the opposite. If your buddy is playing blackjack at the card table and takes a hit (an extra card) when he has 19 and the dealer is showing 4, you should call him a moron. The statistics tell you to stick (decline a card) because the most probable event is that the dealer will bust (get more than 21) or have less than 19. If your buddy takes a card anyway and gets a 2, giving him 21, and wins, should he be hailed as a genius? No, he's still a moron—just a lucky moron. The same holds for any decision we make in the face of uncertainty. Luck doesn't make us smarter or dumber, only lucky or unlucky.

The very next week the Patriots hosted their division rivals, the New York Jets, who had beaten the Pats a few weeks earlier. On their second drive, New England faced fourth down and one on the Jets' 38-yard line. Despite the beating he'd taken in the media, among fans, and even from former Patriots players, Belichick again went for it, which is exactly what the numbers tell you to do. In the broadcast booth, the announcers were leery, already questioning the coach's tactics, "especially after what happened the previous week!" they intoned. This time, however, Laurence Maroney, the Pats' bruising running back, busted over the left tackle for two yards. First down. The announcers said little. Belichick was not praised for this strategic success commensurately with how he'd been blasted the previous week.

Again, this is Bill Belichick. If the most secure coach in the league, whose cerebral analysis is thought to be unmatched, could be subjected to such a severe beating over a well-calculated risk, imagine how a rookie coach or a coach on the hot seat is going to be treated.

And it's not just football coaches who face a difficult time departing from convention. In 1993, Tony La Russa was managing the Oakland A's and was dismayed as his team was last in the division. Pitching was particularly problematic. Oakland's earned run average (ERA) had swollen to more than 5.00. After a particularly brutal weekend series during which the A's gave up 32 runs, La Russa and his longtime pitching coach, Dave Duncan, asked themselves, "Who made the rule that teams need four starters who throw 100 or so pitches, followed by a middle reliever and a closer?"

La Russa seized on an idea: Why not take his nine pitchers and establish three-man pitching "units" in which each pitcher would throw only 50 tosses, usually within three innings? The thinking was simple: The pitchers would take the mound every three games but would be fresher since they'd throw fewer pitches per outing. Also, the opposing batters would be unable to establish much comfort, since they might well face a different pitcher every time they came to the plate. It turns out that baseball statistics back this up. Major League batters hit about 27 points lower the first time they face a pitcher in a game. Their on-base percentage is about 27 points lower and their slugging percentage is 58 points lower the first time they face a pitcher. This could be because the pitcher's arm is fresher or because the hitter needs to see him more than once to figure him out. Either way, La Russa's idea would capitalize on this effect.

There were other potential advantages, too. By having essentially all your pitchers available to you each game, you have more options to choose from in any situation. In addition, the most expensive pitchers tend to be starters who go deep into the game, pitching seven or more innings and throwing 120-plus pitches per game. Turns out the key difference between star pitchers and other pitchers is the stars' ability to pitch effectively for longer. In the first couple of innings, the differences between star and non-star pitchers are much smaller. In La Russa's experiment, for the first three innings he might get comparably effective results from journeyman pitchers who came at a fraction of the cost of the star pitchers, thus leaving extra money to spend on other players—or, in the case of the Oakland A's, allowing them to remain competitive despite a much smaller budget than some of the big-market teams, such as the New York Yankees.

It was a radical strategy, but La Russa had the status and standing to try to pull it off. He'd been the Oakland manager since 1986 and had taken the team to the World Series in 1988, 1989, and 1990. In 1992, the previous season, he had been named manager of the year. With his accumulated goodwill (and his team in last place), he wasn't risking much by departing from conventional wisdom.

Unfortunately for La Russa, his chemistry experiment fizzled. Why? The starting pitchers hated it. Publicly they claimed they had a hard time finding a rhythm and settling into a groove. Privately they complained that the 50-pitch limit precluded them from working the requisite five innings to get a win, yet they were still eligible for a loss. (Because future contracts were tied to wins and losses, their manager was potentially costing them real money.) After five games, four of them losses, and a lot of grumbling from the pitchers, La Russa cut bait and returned to the traditional four-man, deeper-pitch-count rotation. It was a reminder: You may have a better strategy, but if the athletes don't buy in, it's probably not worth deploying.

Here is a cautionary tale of what happens to a risk-taking coach on shaky employment footing. Paul Westhead, coach of the Los Angeles Lakers, was fired 11 games into the 1981–1982 season, in part because the team's point guard, Magic Johnson, thought the coach was, of all things, too rigid and restrictive. "This team is not as exciting as it should be," the Lakers' owner, Dr. Jerry Buss, said at Westhead's firing. By the end of the eighties, Westhead, a Shakespeare scholar who looked the part of a professor, was coaching at the college level, at Loyola Marymount. There he deployed a strategy based on many of the same principles that Kevin Kelley uses in Arkansas: The more offensive opportunities and attempts, the better. The statistics support attempting lots of "big plays"—three-pointers in basketball. The unconventional approach upsets the opponents' preparation routines and displaces them from their comfort zone.

In the 1989–1990 season, tiny Loyola Marymount was the toast of college basketball, the up-tempo team averaging a whopping 122 points a game, running other teams to exhaustion, and coming within a game of reaching the Final Four. (That the team's star player, Hank Gathers, died during the season added a sad layer of drama and exposure.)

Intrigued by Westhead's unique philosophy, his willingness to take ordinary "running and gunning" to a new level, the NBA's Denver Nuggets poached him from the college game to be head coach for the 1990–1991 season. He stated that his methods would be even more effective a mile above sea level, as opponents would tire even more quickly. Westhead encouraged his players to play at a breakneck pace, shoot once every seven seconds—twice the league average—and take plenty of three-pointers. He reckoned that not only would shooting 35 percent on three-pointers yield more points than shooting 50 percent on two-pointers, but longer shots would lead to more offensive rebounds: When the Nuggets missed, they stood a better chance of retaining possession. On defense, the team played at the same methamphetaminic speed, using constant backcourt pressure and trapping. "The idea is to play ultrafast on offense and ultrafast on defense, so it becomes a double hit," Westhead explained to *Sports Illustrated*. "And when it works, it's not like one and one is two. It's like one and one is seven."

Except that it wasn't. At the pro level, Westhead's experiment failed spectacularly. Opposing players took advantage of the Nuggets' chaos and the irregular spacing. The Nuggets' strategy of shooting early and often led to easy baskets on the other end. As it turned out, it was the Denver players who were often huffing and puffing—and on injured reserve—from the relentless running. (One Denver player complained that his arm hurt from throwing so many outlet passes.) Games came to resemble the Harlem Globetrotters clowning on the Washington Generals. In one game, the Phoenix Suns scored 107 points, most on dunks and layups, in the *first half,* which still stands as an NBA record. The Nuggets started the season 1–14 and finished a league-worst 20–62. They scored 120 points a game but surrendered more than 130 and were mocked as the Enver Nuggets, a nod to their absence of "D." Westhead grudgingly slowed down the pace the next season but was fired nevertheless.

You might say it was a valiant effort by Westhead. Hey, at least he tried something different. And if his nonconformist ways failed in Denver, they sure worked at Loyola Marymount. Maybe it was just a question of personnel and circumstance. Barely a decade later, the Phoenix Suns, blessed with better players than Westhead's Nuggets, were borrowing many of his ideas and principles, racking up wins with a celebrated breakneck, shoot-first-ask-questions-later offense nicknamed "seven seconds or less."

But Westhead was hardly cast as an innovator. He was considered an "eccentric," one of the more damning labels in sports. Mavericks are seldom tolerated in the coaching ranks. A mad professor without tenure, Westhead—unlike so many who fail conventionally—never got another NBA head coaching opportunity. His next job was with a modest college program at George Mason University. From there, he caromed to the Japanese League and the WNBA, where he coached the Phoenix Mercury to a title. He returned briefly to the NBA as an assistant, but that was short-lived. At this writing, Westhead is the head women's basketball coach at the University of Oregon, coaching a mediocre team that scores prolifically.

■ ■ ■

Pulaski's Kevin Kelley is an innovative thinker, but he is also exquisitely well placed to install his unconventional strategies. In addition to coaching the football team, Kelley doubles as the athletic director for Pulaski Academy. He is his own immediate supervisor. He draws his players from a small pool of affluent kids whose parents can afford parochial school tuition and probably place football a distant third behind academics and violin lessons. When Kelley's choices fail, there aren't many boos from the stands or angry fans calling the local sports talk show or starting websites dedicated to his firing. Since he coaches high school kids, he doesn't face the threat of player (and agent) revolt the way Tony La Russa did with the Oakland A's pitching staff.

That Thursday night game at Pulaski spanned nearly three hours, mostly because of incomplete passes and penalties that stopped the game clock. But it showcased how Kelley's savvy and well-considered, if unconventional, approach led a decidedly smaller, slower, and younger Pulaski team to victory, 33–20. Afterward, in the postgame breakdown, Kelley said flatly, "The system won that game." As the players shook hands near midfield, one of the Mustangs sought out the Bruins' quarterback, Wil Nicks, and told him, "I wish we played like y'all."

Pulaski went all the way to the Class 5A state championship game in 2008. In that tournament run, Kelley stayed true to his philosophy. In the semifinal game against Greenwood—the school that had knocked them out of the tournament two years in a row, including a 56–55 heartbreaker in the state championship in 2006—Pulaski started the game with an onside kick, recovered it, and drove all the way down to Greenwood's six-yard line before turning the ball over after failing to convert on fourth down. That might have discouraged most coaches, especially against a team they've had trouble beating. Not Kelley. He continued to go for it on every fourth down, eventually winning the game 54–24 and amassing 747 yards of total offense in the process.

In the championship game against West Helena Central—a team with eight future Division I players to Pulaski's one—Kelley again refused to punt or kick. In the waning minutes, the Bruins had possession and clung to a slim 35–32 lead. Faced with three fourth downs early in the drive, they went for it each time and made it. With less than 1:30 left on the clock, they faced yet another fourth down at midfield. The conventional strategy was to punt the ball, pin your opponent deep in their own end, and force them to drive 60 to 70 yards in less than a minute and a half to get into field goal range. If you go for it and fail, you leave Helena just 20 yards away from field goal range and give them a chance to tie the game. What do the statistics tell you to do? Go for it. That is what Kelley did. The Pulaski quarterback plunged over the right side for a couple of yards, converting yet another fourth down on what would be the final drive of the game as Pulaski ran out the clock and captured its second state championship. Asked if he ever thought about punting on that final drive with so much at stake, Kelley responded without hesitation: *"Never."*

For kindred spirits in the coaching ranks who are tempted to topple conventional sports wisdom, Kelley has the same advice he gives his teams on fourth down: Go for it. Until they do, at least players have a response at the ready the next time their coaches accuse them of being soft or making boneheaded decisions or failing to do everything they can to help the team win. "Sorry, Coach, but I'm just following the example you set with your play-calling."

HOW COMPETITIVE ARE COMPETITIVE SPORTS?

Why are the Pittsburgh Steelers so successful and the Pittsburgh Pirates so unsuccessful?

The noise level and the sun rose in tandem. A couple of nights earlier the New York Yankees had won the 2009 World Series, and now, on this chilly November morning, it was time for their parade. Fans had been lining the streets of lower Manhattan since the infomercial hours. By 7:00 A.M. the crowd was five deep. An hour later the inevitable "Let's go Yankees, tap-tap-taptaptap" cheers began. Kids pulled from school sat regally on their parents' shoulders. The New York cops, their spirits buoyed by the overtime they were racking up, were uncommonly friendly. Wall Street traders and analysts and bankers peered from their offices overhead and smiled for one of the few times all year. The motorcade wouldn't crawl past until noon, but in a congenitally impatient city where no self-respecting pedestrian waits for the light to change, this was the rare occasion when millions of New Yorkers stood happily along Broadway for hours.

The 2009 World Series parade attracted more than 3 million fans—a greater mass of humanity than the entire *market* of some MLB teams. Among the crowd: former mayor Rudy Giuliani, Spike Lee, and Jay-Z, who performed the civic anthem at the time,

"Empire State of Mind." There were the obligatory keys to the city, mayoral proclamations, and a forest's worth of confetti. It was a tidy snapshot of why the Yankees might be the most polarizing team in all of sports. While the rest of the country seethed and cursed the arrogance and excess, Yankee Nation gloated over still another World Series triumph, the twenty-seventh in the franchise's storied history. Mocking, of course, the milk ad campaign, one T-shirt sold at the parade tauntingly asked of other teams' fans: "Got Rings?"

The answer was probably "no," or at least "not many." The World Series has been held since the early years of the twentieth century, yet only a few franchises have won a significant number of titles. Eight current organizations have never won a World Series, and nine others have won fewer than three. The Texas Rangers have been around in one form or another since 1961, and prior to 2010 they had never even *been* to the Fall Classic, much less won it. In contrast, since 1923, the Yankees have won on average once every three years.

As a rule, we're offended by oligopolies and monopolies. We much prefer competition; it's healthier, it's better for consumers, it encourages innovation, it just feels fundamentally fairer. We have antitrust laws to promote competition. We're careful to crack down on cartels and regulate industries—yes, some more than others. In the heavily regulated airline industry, the largest carriers in the domestic market, American and Southwest, each have less than 14 percent of the market share. Banks, too, are heavily regulated, so much so that under the so-called Volcker Rule, no institution may exceed a 10 percent market share. Citigroup may have been deemed "too big to fail," but its market share is only 3 percent. Walmart might be the American company most maligned as a monopoly, but in 2009 its share of the $3 trillion U.S. retail market was 11.3 percent.

The Yankees? Inasmuch as World Series rings constitute a market, their market share is 25 percent.

How can one team dominate like this while other teams are barely competitive? The quick and easy answer is money, especially in the absence of a salary cap. Fans of 29 other teams will note that when the Yankees can spend north of $200 million on players, as they did in 2009, and most other teams spend less than $100 million, they're naturally going to have a heavy concentration of titles. They'll handily beat the Phillies—their opponents in that World Series—who spent "only" $113 million on payroll. Just as in the previous year, the Phillies ($98 million) beat the Tampa Bay Rays ($44 million), and the year before that the Boston Red Sox ($143 million) beat the Colorado Rockies ($54 million). No wonder the small-market Pittsburgh Pirates—2010 payroll, $39.1 million—haven't had a winning season since 1992.

However, the reason for the Yankees' extraordinary success is more complex than that. Just about everything in baseball's structure militates against parity. Start with the 162-game season. In the same way an opinion poll sampling 100 subjects will be a more precise reflection of the way the public thinks than a poll sampling 10 subjects, baseball's long season lends itself to an accurate reflection of talent. If two teams play one game, anything can happen, but if they play a good many games, the better team will win the majority of the time.

Then consider the playoffs. Only the eight best teams make it to the playoffs, so 22 are out of the running. Teams play a best-of-five-game series followed by a best-of-seven League Championship Series followed by a best-of-seven World Series. As with the regular season, the sample size is large enough that the best team ought to win the series, especially with a home field advantage. The Yankees may *be* the best team in baseball because they buy the best players, but the imbalance is allowed to flourish because of baseball itself.

Contrast this with the NFL, the league that openly strives for parity and democracy. The season spans only 16 games, hardly a robust sample size. A few breaks or injuries could represent the difference between a 7–9 season and a 9–7 season. Not only do 12 teams qualify for the playoffs, but there is no "series format." It's single elimination, "one and done," a format much more conducive to upsets, much more likely to generate randomness. One unlucky game, one untimely injury to a star player, and it's easy for a lesser team to win and move on. Plus, until 2010 there was a salary cap that prevented the wealthy teams or the teams blessed with cavernously pocketed owners from outspending their rivals by factors of three and four. And with the bulk of team revenue coming from leaguewide television contracts, the schism between the economic haves and have-nots is much narrower than in baseball.

The result? As you'd expect, the concentration of champions is lowest in football, the "market share" remarkably balanced. The NFL has been holding the Super Bowl only since 1967, but already 18 of the 32 franchises have won the Lombardi Trophy and all but 4 have appeared in the Super Bowl at least once. (That's almost the same number of teams that have never been to the World Series—and they've been holding that since 1903.) Market size doesn't matter much, either. Most Super Bowls? The Steelers, with six, hailing from . . . Pittsburgh, the same town that hasn't fielded a competitive baseball team in almost 20 years. The Packers from Green Bay, Wisconsin, the smallest market in major U.S. professional sports, have won three titles.

There are far more than 16 games in the NBA and NHL regular seasons, and the playoffs are seven-game series. That cuts against randomness and in favor of the monopolies. In contrast, unlike in baseball, more than *half* the teams make the postseason. And the NBA and NHL both have a salary cap. So we shouldn't be surprised to learn that the concentration of champions in pro basketball and hockey is significantly greater than in the NFL and significantly less than in Major League Baseball.

Three months after the World Series parade in New York there was a similar processional for the Super Bowl champs in small-market New Orleans. The city sported a few hundred thousand fans rather than a few million. And this wasn't the franchise's twenty-seventh title; it was the first. But it was just as jubilant. A week before Fat Tuesday, players rode around on floats, wearing masks and tossing beads. Lombardi Gras, they called it.

Trying to predict who will win the next Super Bowl is a fool's errand, but trying to predict who will win the next World Series is far easier. Though you might not be right, you can limit your potential candidates to a handful of teams even before the season begins. Funny thing about sports: Distilled to their essence, they're all about competition. But as an industry, some are more competitive than others.

TIGER WOODS IS HUMAN

(AND NOT FOR THE REASON YOU THINK)

How Tiger Woods is just like the rest of us,
even when it comes to playing golf

It started with his father. In one of the great money quotes in the annals of sports, Earl Woods confided to *Sports Illustrated* that his son, Tiger, not merely would transcend golf, or sports, or even race but would transcend civilization. "Tiger will do more than any man in history to change the course of humanity," Earl said without a trace of irony. "He's qualified . . . to accomplish miracles. He's the bridge between the East and the West. There is no limit because he has the guidance. I don't know yet exactly what form this will take. But he is the Chosen One. He'll have the power to impact nations. Not people. *Nations.* The world is just getting a taste of his power." The year was 1996, and Tiger, age 20 at the time, had yet to win his first Major title.

Instead of dismissing these claims as the messianic ranting of another crazily ambitious sports parent—if Tiger was Jesus, what did that make Earl?—many actually stopped to consider the prophecy. Could it be that Old Man Woods had it right? In the years that followed, Tiger did little to discredit his father's prediction.

By his mid-twenties, Tiger had single-handedly hijacked professional golf and, with the retirement of Michael Jordan, was on his way to becoming the brightest star in the entire sports cosmos.

When Tiger played, he usually won. When he didn't play, events had the thrill of Christmas without Santa. At this writing, he's won 14 Major titles for his gilded career and, despite a recent slide, still is a good bet to eclipse Jack Nicklaus's record of 18. Through 2009, Tiger had won roughly 30 percent of the events he'd entered. For the sake of comparison, Nicklaus won 73 titles in 594 events, or 12.3 percent.

It's not just the relentless winning that has perpetuated the Tiger-as-Chosen-One mythology. It's *how* he has won. His game was unparalleled. His physical gifts were matched by his neurological gifts. He was the best at driving and the best at putting. He blew the field away; he trailed and then rallied on Sunday. He won with wise and conservative play; he won with brazen, you-must-be-kidding-me shot-making. He prevailed at the 2008 U.S. Open playing on what we later learned was a shredded knee. He performed miracles such as the famous chip shot on the sixteenth hole at the 2005 Masters, an absurd piece of handiwork that defied all prevailing laws of geometry and physics.

Though Earl Woods passed away in 2006, over the years others joined his "Messiah chorus." *Esquire* magazine described Tiger as "Yahweh with a short game." The commentator Johnny Miller once declared, "Like Moses, [Tiger] parted the Red Sea and everyone else just drowned." Even Woods's mother, Kultida, seemed to buy in, at one point remarking: "He can hold everyone together. He is the Universal Child." Inevitably perhaps, a website, tigerwoodsisgod.com, "celebrating the emergence of the true messiah," came into being. (The site's "Ten Tiger Commandments" include the directive "Thou shalt not covet Tiger's game.")

Charles Barkley, always good for the unvarnished truth, declared that other PGA Tour players "are afraid of black Jesus." Perhaps because it's easier to rationalize the ritual butt kickings, check out how other golfers characterize Woods. "He is something supernatural," declared Tom Watson. "He is superhuman,"

asserted Paul Azinger. The apotheosis, so to speak, of Tiger dei-
fication came a few years ago when EA Sports made a golf video
game. On one of the holes, Woods removes his Nikes, rolls up his
slacks, and walks into a pond to hit a shot that naturally lands in
the cup. It's known as the Tiger Woods Jesus Shot.

On Thanksgiving night in 2009, Woods was injured in what
was first described as a "car accident." The car accident quickly
morphed into a train wreck, a sensational sex scandal that, perhaps you've heard, linked Woods to an unceasing string of
women—porn stars, diner hostesses, reality show rejects—none
of them his wife. Apart from the sheer tawdriness, the scandal
had "legs" because of Tiger's starring role. Headlines the likes of
"Tiger's Harem Grows" were jarringly at odds with a figure perceived as immortal. As Woods sought to assure us in his first public statement: "I'm human and I'm not perfect."*

But even before the scandal there existed conclusive proof that
Tiger Woods is in fact mortal. What's more, this proof comes from
the way he golfs. Tiger putts the same way you and I do. He's
immeasurably more accurate, fluid, and poised, and his scores
are much lower. But at least Tiger is subject to the same faulty
thought process as we are.

Recall loss aversion, the principle that we dislike losing a dollar more than we enjoy earning a dollar. As a result of loss aversion, we change our behavior—sometimes irrationally—paying
too much attention to purchase price and avoiding short-term loss
at the expense of long-term gain. In theory, tax purposes notwithstanding, what we paid for something is irrelevant. All that should
matter is what it's worth today and what it will be worth in the
future. But we don't behave that way. Investors routinely sell winning stocks too early and hold on to the dogs for too long. Home
owners often do everything in their power to avoid selling their
property for less than the purchase price. Texas hold 'em players

---

* In the wake of the scandal, there's now a companion website, www.tiger
woodswasgod.com.

depart from their strategy when their stacks of chips diminish. And golfers, even the pros, neglect their overall score to avoid a loss on a single hole.

How do we know this? A few years ago, two professors, then at Wharton, Devin Pope and Maurice Schweitzer, began looking at the putting tendencies among 421 golfers on the PGA Tour in more than 230 tournaments. Using over 2.5 million laser-measured putts from tour events held between 2004 and 2009, they measured the success rate of nearly identical putts for birdie, par, and bogey. The idea was simple. Each hole on a golf course has a "par" score—the number of strokes you're expected to take before depositing the ball in the hole. A shot in excess of par is, of course, a bogey. One shot less than par is a birdie. Put another way: A bogey is a "loss" and a birdie is a "gain" on that hole.

In golf, however, the only measurement of true significance is the *total score* at the end of all 18 holes, so a player shouldn't care whether he is putting for a birdie or par or a bogey on a hole. The idea is to maneuver the ball into the hole in as few strokes as possible on *every hole, no matter what*. Analogize this to your retirement portfolio. You simply want the most favorable total at the end. It shouldn't matter how you got there.

The study, however, found something peculiar. When a golfer on the PGA Tour tries to make a birdie, he is less successful than when he lines up the *exact same putt* for par. The researchers were careful to measure the exact same distance (accurate to within a centimeter) of each putt, from the exact same location on the green, and from the exact same hole. In other words, they were looking at literally the *same* putt for birdie versus par from the same location on the green on the same hole. Even Tiger Woods—so unflappable, so mentally impregnable—changes his behavior based on the situation and putts appreciably better for par than he does for a birdie, evaluating decisions in the short term rather than in the aggregate.

The explanation? The same loss aversion that affects Wall Street investors, home sellers, and consumers informs putting on the PGA Tour. Professional golfers are so concerned with a loss that they are more aggressive in avoiding a bogey than they are in scoring a birdie. Remember the dieters who weren't motivated to lose weight until they faced the possibility of paying a $1,000 fine? Golfers operate the same way. Dangle the "bonus" of a birdie—the gain of a stroke—and it's all well and good. Says Pope, "It's as if they say, *Let's get this close to the hole and see what happens*." But threatened with the "deduction" of a bogey—the loss of a stroke—they summon their best effort. "They're telling themselves," says Pope, "*This one I gotta make*."

The professors also found something interesting to confirm the more aggressive behavior on par versus birdie putts. When professional golfers missed their putts for a birdie, they tended to leave the ball disproportionately short rather than long. This was evidence of their conservative approach. They were content to set up an easy par by leaving it short and not risk overshooting, which might leave a more difficult putt for par. When the same putts for par were missed, it wasn't because they fell short.

The two researchers also tried to rule out all other potential explanations by controlling for the day of the tournament, how far off a golfer was from the leader board, how the previous holes were played, and what number hole was being played. None of those factors changed the tendency to putt differently for par than for a birdie.

The authors estimated that when the average golfer overvalues individual holes at the expense of his overall score, it costs him one stroke for each 72-hole tournament he enters. That may not seem like much, but most golfers would kill to improve their game by one stroke. For a top golfer like Woods, this mismanaged risk has the potential to cost him more than $1 million in prize money each year.

Tiger even appears to be aware of his loss aversion. As he told the *New York Times*, "Anytime you make big par putts, I think it's more important to make those than birdie putts. You don't ever want to drop a shot. The psychological difference between dropping a shot and making a birdie, I just think it's bigger to make a par putt."

It's somehow reassuring that Tiger is, at least in this respect, decidedly human. However, Pope has a point when he says: "If Tiger Woods is biased when he plays golf, what hope do the rest of us have?"

This was thrown into sharp relief at the 2009 PGA Championship at the Hazeltine National Golf Club in Minnesota. Heading into the final round, Woods appeared to be cruising inexorably to still another Major championship. It wasn't just that he was carving up the course and leading the pack, eight shots under par. His unlikely challenger, Yang Yong-eun, Americanized to Y. E. Yang, was unknown even to hard-core golf fans.

Yang's anonymity was such that television researchers and media members scrambled to find basic biographical info. It turned out he was the son of South Korean rice farmers and didn't discover golf until he was 19. Until then, Yang had been an aspiring bodybuilder, but he injured his knee and channeled his frustrations at a local driving range. Teaching himself golf mostly by watching instructional videos, Yang was able to break par by his twenty-second birthday. Unfortunately, that was also the year he was required to show up for mandatory military duty. When his service ended, he returned to golf and slowly worked his way up the sports org chart, from the Korean regional tour to the Asian tour to qualifying school, eventually earning his card on the PGA Tour.

Heading into 2009, Yang, then 37, was making a living but not much more. He had never won a PGA event and had posted only one top-ten finish. Even at Hazeltine, he was lucky to make the cut after shooting a shaky 73 in the first round. (Tiger had shot a 67.) Now here he was in a showdown against Tiger Woods for a Major. Even Yang admitted that his overarching goal was to not embarrass himself. "My heart nearly exploded from being so nervous," he recalled.

But under the principle of loss aversion, in the face of loss, we perform more aggressively. Sure enough, facing an almost certain loss, Yang could let it rip and play with devil-may-care abandon. Woods, by contrast, was facing an almost certain gain—a lead, an inexperienced challenger, and, above all perhaps, a 14-for-14 record of closing out Majors when leading after 54 holes. But in the face of gain, we perform conservatively, more concerned about "don't-mess-this-up" defense than about "gotta-get-it-done" offense. In essence, the entire round was one big birdie putt for Woods.

You may remember the remarkable outcome. Playing with a striking absence of aggression, Woods shot three shots above par, a score of 75 that included an astounding 33 putts. "I felt that with my lead, I erred on the side of caution most of the time," Woods conceded. Yang, in contrast, played with what *Sports Illustrated* called "carefree alacrity." He smiled, shrugged, and went for the pin every time. A stroke ahead on the eighteenth hole, Yang continued to play with a level of audacity that suggested that he still believed he was facing a loss. After a solid drive, he was 210 yards from the hole. On the approach, he used his hybrid club to try to loft the ball over a tree and onto the green. It was a shot as intrepid as it was difficult. And Yang nailed it, maneuvering the ball within eight feet of the hole. He putted out for a sensational birdie and won the tournament, becoming the first Asian to capture a Major championship in golf. And with the help of loss aversion, he'd humanized Tiger Woods.

■ ■ ■

Not that golfers are unique. Look closely and you'll see that virtually all athletes, just like the rest of us, are affected by loss aversion in one form or another. Imagine a pitcher jumping to an 0–2 count on a batter. The pitcher is probably thinking, *I'm gonna get a strikeout here,* or, at the very least, *I'm gonna get this guy out,* since he's far ahead in the count. The pitcher has already accounted for the "gain" of an out. Then, after a few more pitches, the count is 3–2. Suddenly the pitcher is in danger of losing what he thought he had.

Now imagine the same pitcher in a different situation. He throws three lousy pitches to the batter, and the count is 3–0. He's likely to think: *Damn, I'm gonna walk this guy.* But then he steadies himself and throws a pair of strikes, or perhaps the batter fouls off two pitches, or the umpire gives him a couple of favorable calls. The count is now 3–2. Suddenly the prospect of an out takes on the dimension of an unexpected bonus.

In a vacuum, the pitcher should handle the two situations identically, right? In both cases the count is 3–2, and how he arrived there—i.e., the purchase price—shouldn't matter. The goal is simply to get the batter out, much as the goal of the golfer is to accumulate the lowest cumulative score over 18 holes or the goal of the retiree is to accumulate the fattest retirement portfolio. Intuitively, we might expect the pitcher starting 0–2—perhaps questioning his control after throwing three straight balls—to throw conservatively and the pitcher who started 3–0 but has thrown two straight strikes to be more aggressive.

But that doesn't account for loss aversion.

Inspired by the golf study, we looked at three years of MLB Pitch f/x data (more than 2.5 million pitches) and accumulated all 3–2 counts that started off as either an 0–2 count (where the pitcher is now staring at a potential short-term loss: the loss of the out he thought he had) or a 3–0 count (where the pitcher is facing a short-term gain). We then examined how the pitcher threw the next pitch. We found that when pitchers face a 3–2 count that started off 0–2, they throw far fewer fastballs and more changeups and curveballs than do pitchers facing the same full count but who started off 3–0. In a full count, a pitcher who starts off 0–2 is 51.5 percent likely to throw a fastball, 21.0 percent likely to throw a curve, and 8.2 percent likely to use a changeup. The same pitcher facing the same 3–2 count who starts off 3–0 throws a fastball 55.4 percent, a curve 17.7 percent, and a changeup 7.3 percent of the time.

This is consistent with the principles of loss aversion. Changeups and curveballs are more risky and aggressive pitches. Pitchers will tell you that fastballs are more reliable and conservative. So a pitcher facing the possibility of a loss, because he used to be ahead in the count, will throw more aggressive pitches to avoid that loss. "I gotta get this guy out now!" The same pitcher in the same situation who was once behind in the count will throw more conservatively. He feels less urgency, and his best effort is not being summoned as a result. Like the golfer attempting a conservative putt for birdie, his choice suggests the attitude "I didn't expect to be here anyway, so no great loss if this doesn't work out."

Even more interesting: Pitchers facing a mental loss because they were once ahead in the count 0–2 not only pitch more aggressively but achieve more favorable outcomes. They're more likely to strike out the batter—from swinging and missing as well as from called strikes. In addition, batters in these situations are less likely to make contact with the ball. They foul off fewer pitches and put the ball in play less often—and when they do put it in play, it's more likely to result in an out.

The batting average of Major League hitters facing a pitcher with a 3–2 count who was once ahead 0–2 is only .220 compared with a batting average of .231 when facing the same 3–2 count against the same pitcher who once was behind in the count 3–0. That's an 11-point difference for the same count against the same pitcher. Slugging percentages are nearly 20 points lower (0.364 versus 0.382), and virtually all other hitting statistics are lower in these situations as well. As with the higher success rate for identical par versus birdie putts in professional golf, pitchers who adopt *less* conservative strategies because of loss aversion fare better.

We can also look at this from the batter's perspective. A loss to a pitcher is a gain to a batter. Thus, a batter facing a 3–2 count who was initially in an 0–2 hole views this as a mental gain: "I thought I was going to strike out, but now I could easily walk or get a hit." And a batter who was initially up 3–0 views the full count as a potential loss. "I thought I was going to reach base, and now I might not." Loss aversion predicts that the batter will behave more aggressively in full counts when the count was previously 3–0 and more conservatively when the count was previously 0–2—the opposite behavior of pitchers. And that's true.

Batters are more conservative on 3–2 counts if they started out 0–2, swinging at fewer pitches, even those down the center of the strike zone. And when they swing, the outcomes are worse: more strikeouts, fewer balls put in play, and when they are put in play, more outs.

Thus, it's no surprise that their hitting numbers are lower—batters become too conservative at precisely the time when pitchers are becoming more aggressive. Similarly, a batter who was previously ahead 3–0 in the count will be much more aggressive on a 3–2 pitch, just when the pitcher becomes more conservative. Considering this behavior, the difference in hitting statistics makes a lot more sense.

■ ■ ■

In football we can conduct a similar field experiment (literally). Ask yourself, when are teams in identical situations more likely to go for it on fourth down based on where they started the series? Let's imagine that two teams each have the ball fourth and goal at the one-yard line. In the first example, the team started the series on the one-yard line and in three unsuccessful plays did not move the ball. In the second example, the team started at the ten and gained nine yards in three plays.

In the first example, you've seen that either the other team's goal-line defense is really good or the first team is having a hell of a time moving the ball 36 inches. *Why tempt fate?* Kick the field goal, right? In the second example, they've moved the ball nine yards in three downs. Odds are good that they can pick up one more yard on fourth down, so they'd be more inclined to go for it, right?

Wrong. The results among NFL teams run *completely counter* to this. Teams in the first example are far more likely to go for it than are those in the second example. Why? Loss aversion. The team that started the drive first and goal at the one-yard line is thinking touchdown. They've already mentally accounted for the seven points. If, a few plays later, it's still fourth and goal, they don't want to lose the touchdown they thought was "in the bank."

In the second situation, the prospect of a touchdown is more of a "gain," and the team is more likely to play conservatively, the same way a golfer guides a birdie putt or a pitcher throws a fastball on a 3–2 count when he started off 3–0.

Facing fourth and goal from the one-yard line, NFL teams go for it 67 percent of the time if they started with first and goal from the one-yard line but only 59 percent of the time if they started first and goal from the ten-yard line. More generally, if teams that are facing fourth and goal from the one started inside the three-yard line, they go for it 66 percent of the time. But if the same teams started from the eight-yard line or farther out, they go for it only 61.5 percent of the time. This is exactly the opposite of what many would expect.

Loss aversion is a powerful tool for predicting when teams will go for it on fourth and goal. When a team starts out first and goal at the one-yard line and is then *pushed back* to fourth and goal at the two- or three-yard line, the likelihood that they'll go for it is 35 percent. And if it's fourth and goal from the two- or three-yard line and they didn't start out at the one? They go for it only 22 percent of the time. In other words, even when pushed back a couple of yards—implying that the defense is making a strong goal-line stand or that the offense has been ineffective—teams are still much more willing to go for it than if they had been moving the ball forward and found themselves in the same position. Exactly the opposite of what most of us might expect, but consistent with loss aversion.

Here's another way to evaluate the power of loss-averse behavior in the NFL. An extreme case of shortsighted loss occurs when a team scores a touchdown that is then nullified by a penalty. Imagine a kickoff for a touchdown. The returner makes a sharp cut, sees an empty field before him: 50, 40, 30, 20, 10. . . . He crosses the goal line, spikes the ball in the end zone, and is mobbed by teammates while the coach high-fives his assistants. But wait, there's a flag on the play: an illegal block. So the team starts the drive back on its own 20-yard line.

How do teams respond, having gone from the ecstasy of gain to the agony of loss so quickly in such situations? On drives in which a touchdown was called back because of a penalty, teams are 29 percent more likely to go for it on fourth down than they would have been otherwise (controlling for the number of yards to go, the position on the field, and the score in the game). Loss aversion dictates that the team will fight like crazy to get that touchdown back. And teams attempt to do so on *that drive,* as opposed to later in the game. Of course, whether a team scores on any particular drive is largely irrelevant. All that matters is the final score.

Loss aversion affects the NBA in a similar way. Team A is winning by a healthy margin and probably is thinking, "We've got this game in the bag." Mentally, they've already chalked one up in the win column. Then Team B makes a comeback, as NBA teams so often do. Suddenly the win Team A thought it had is in doubt. Worse, if Team B actually takes the lead heading into the fourth quarter, Team A is facing a potential loss. Conventional wisdom suggests that Team A will play passively. Countless times, we've heard a losing coach in this situation complain, "We stopped being aggressive." Yet the principles of loss aversion suggest that in the face of this kind of loss, the team will play *more* aggressively. It is in the face of a gain that they will play more conservatively. Who's right?

Examining nearly 5,000 NBA games, we studied situations in which two teams headed into the fourth quarter within 5 points of each other but one team had led by at least 15 points in the third quarter. In other words, we looked at the final 12 minutes of close games in which one of the teams came from behind by a significant margin. We then subdivided our sample into two scenarios: In the first, the team that was ahead by 15 or more is still ahead, but by fewer than 5 points. This team is still facing a gain, but the prospect of a win is no longer as certain as it once was. In the second situation, the team that was ahead is now down by fewer than five points heading into the final period. Here, that team is facing a gut-wrenching loss. Its lead has evaporated, and now it's behind, going from what was a sure win to the real possibility of a loss.

It turns out that teams that had once been ahead by a lot but are now trailing by a few points in the fourth quarter start to play very aggressively: They shoot more three-pointers and shoot more frequently, taking shots four to five seconds faster than they normally do. This is exactly what loss aversion predicts. Facing a potential loss in a game they were sure they would win, like golfers facing par putts, they ramp up the aggression. By contrast, the team that previously had a large lead and is now up only a few points at the beginning of the fourth quarter starts to play very conservatively: Its players shoot fewer three-pointers and shoot less frequently, taking more time than normal between shots (i.e., holding the ball longer).

Time and again, we hear coaches implore players, "Forget about what just happened," "You can't change the past," or "Put it behind you." The message: It doesn't matter how you arrived at this point, just play as you normally do. In theory, they're right. But it's like asking the home owner to forget about her purchase price when she considers a lower offer on her property. For professional athletes, the past is relevant and it's hard to block out *how* they got into their current predicament.

Research by Antonio Damasio of USC and George Loewenstein of Carnegie Mellon laid bare the power of loss aversion with a curious experiment. They revisited classic loss aversion experiments but tested subjects with brain damage in the area that is thought to control emotion. Compared with normal subjects, the emotion-impaired patients did not exhibit the same penchant for loss avoidance. As a result, in an investment game the researchers had designed, the brain-damaged, emotionally impaired subjects significantly outperformed the other, normal subjects. Why? Because they treated losses no differently from gains. The lesson? Short of a lobotomy, we all fall victim to loss aversion.

Loss aversion influences everything from everyday decisions to athletic performance to individual investments. It also affects our behavior as sports fans. Thanks to loss aversion, we tend to place a higher value on objects we *own* than on objects we don't even if it's the same object. In theory, our willingness to pay for something should be the same as our willingness to be deprived of it.

If you value LeBron James at $40 in your fantasy league, presumably you would pay $40 to own him or accept $40 to sell him. But it seldom plays out this way.

This phenomenon, related to loss aversion, has a name—the endowment effect—coined by Richard Thaler, a behavioral economist at the University of Chicago. Thaler found that people feel the loss of something they own much more deeply than they feel the loss of something they don't own. If we give you $100 and then take it away, that's much more painful than telling you that we were going to give you $100 but decided not to.

To demonstrate the endowment effect, Dan Ariely and Ziv Carmon, two behavioral psychologists at Duke University, performed an experiment using basketball tickets. Duke, of course, has an exceptionally successful basketball team. It also has an exceptionally small basketball arena, the 9,314-seat Cameron Indoor Stadium. For most games, demand for tickets greatly outstrips supply. To allocate seats, the university has developed a complex selection process, and as much as a week before games, fans pitch tents in the grass in front of the arena and wait on line. For certain important games, even those who remain on line aren't guaranteed a ticket, only entry in a raffle.

After tickets had been allocated for a Final Four game, the professors called all the students on the list who'd been in the raffle. Posing as ticket scalpers, they asked those who had not won a ticket to tell them the highest amount they would pay for one. The average answer was $170. When they asked the students who had won a ticket for the lowest amount at which they would sell, the average answer was $2,400. In other words, students who had randomly won the tickets and had them in their possession valued them roughly *14 times* higher than those who hadn't.

For an even more vivid illustration of loss aversion, consider how you, as a fan, respond to wins and losses when your team plays. Your favorite NFL team is winning 30–3, and you're justifiably confident that the game is in the bag. Suddenly the opposition stages a fierce comeback to close the score to 30–27, and you're in panic mode. It turns out that your team hangs on for the win, but you're probably left feeling a bit hollow, less elated and triumphant than relieved and thankful. The other team's fans probably feel disappointed, but it's leavened by the surge of the comeback that fell just short.

Contrast this with what happens when two teams are locked in combat for hours. The lead alternates. Momentum fluctuates. Tension escalates. With the score tied 27–27, your team marches downfield and kicks a game-winning field goal as time expires.

Both games end with the exact same score, 30–27. We're told all the time: "A win is a win is a win." "Winning ugly is still winning." "A blowout doesn't get you extra points in the standings." Again, how your team gets there shouldn't matter, just as the past shouldn't matter when we sell a stock or put a house on the real estate market. But from the perspective of fans, we know that's seldom the case. A last-second field goal to decide a close game? When our team wins, we're doing cartwheels, straining our larynxes, and high-fiving anyone within arm's reach. We're despondent and hurling the sofa cushions at the television when our team loses.

"How we got there" matters because as a game evolves, we adjust our loss-gain expectations accordingly. In the 30–3 game, we *own* the win. We count on it and account for it the same way a team with first and goal at the one-yard line counts on the touchdown. When it's threatened, we face the loss of something we'd assumed was ours. And we hate loss even more than we like gain. Barely hanging on to what's ours when it seemed a lock? Where's the pleasure in that?

In a close game, when we live and die a little with each play, we haven't made an accounting of gains and losses. We were never endowed with a victory; and we never steeled ourselves for a loss. So when the gain comes at the very end, it's ecstasy. Nothing's been unexpectedly taken. And when we lose at the end, we're devastated. It's the same phenomenon that takes hold when you play a Pick 6 lottery game. You chose your numbers, and right away there's no match. Oh, well. You let it go with relatively little emotion. Now imagine that the first five numbers are matches. Only one more to go for a $250 million payoff! The last number comes and . . . it's not a match. Ouch. You lost in both situations. That's all that should ultimately matter, but you feel the loss much more profoundly when the outcome is in doubt right up to the end.

Consider what happened at the annual Yale-Harvard football rivalry—self-aggrandizingly called "The Game"—in 1968. Yale entered the game nationally ranked, brandishing an 8–0 record and a 16-game winning streak. The team's quarterback, Brian Dowling, the biggest of big men on campus, was the figure immortalized as B.D. in the *Doonesbury* cartoon created by younger classmate Garry Trudeau. The lore was that Dowling hadn't lost a game since sixth grade. Yale's other standout was Calvin Hill, a future Dallas Cowboys star running back as well as the future father of basketball star Grant Hill. Harvard also entered the game undefeated. In addition to bragging rights, the winner would take home the Ivy League title.

Yale controlled the game, up 29–13 with less than a minute to play. Yale fans were "endowed" with a gain. Harvard fans girded themselves for a loss. Then the unthinkable happened. After recovering a fumble, Harvard scored an unlikely touchdown. With nothing to lose, it tried a two-point conversion that was successful, making the score 29–21. As everyone in the Harvard Stadium expected, the Crimson attempted an onside kick. Yale fumbled the return, and Harvard recovered at midfield. Already, the anticipated thrill of victory by the Yale fans was being undercut by this flirtation with a loss, just as any agony of defeat by the Harvard faithful would be offset a bit by this late surge.

Harvard methodically moved the ball downfield. On the game's last play from scrimmage, the Harvard quarterback scrambled and desperately chucked the ball to the corner of the end zone. A Harvard receiver snatched the ball for the touchdown. The score was now 29–27. Students were already storming the field when Harvard lined up for a two-point conversion on the final play of the game. The Harvard quarterback knifed a quick pass through the Yale defense that the intended receiver hauled in—29–29. Harvard had scored 16 points in the final 42 seconds—and with no overtime, the game ended in a tie.

The Harvard players, fans, and alumni were, of course, deliriously happy. Yale's were crushed. But wait: The game ended in a tie. Shouldn't both sides have felt an equal measure of pleasure and pain? Yeah, right. The same way a former Lehman Brothers executive once worth nine figures on paper and the newly crowned Powerball winner feel commensurate joy about their respective $5 million nest eggs. How you got there matters.

Forty years after the game, Harvard's Kevin Rafferty, a documentarian, revisited the afternoon and its effects on those involved. He was inspired in part by his father, a Yale football player in the 1940s who fought in World War II but described that Saturday in November 1968 as the "worst day of my life." Many of the players went on to fabulously successful careers in business, law, medicine, and, in the case of Harvard's all–Ivy League tackle Tommy Lee Jones, cinema. But four decades later, memories of that football game for most of them are still fresh, emotions still raw.

The title of the film, pulled from a *Harvard Crimson* headline, neatly summarizes loss aversion: *Harvard Beats Yale 29–29*.

OFFENSE WINS CHAMPIONSHIPS, TOO

Is defense really more important than offense?

The moment had arrived at last. In June 1991, Michael Jordan cemented his reputation as the best player of his era—check that: any era—by leading the Chicago Bulls to the NBA title. As he cradled the trophy for the first time, his explanation for his team's success had a familiar ring to anyone who's ever played team sports. "Defense," Jordan explained, "wins championships." It might have been the most quoted maxim in the sports lexicon, but because Jordan said it, it now had the ring of gospel.

In 1996, the Bulls defeated the Seattle Sonics (R.I.P.) to win the title. By that time Jordan's accumulation of rings had grown to four, yet his analysis of the Bulls' success remained steady. "It's been shown that defense wins championships," he said. A year later, the Bulls beat the Utah Jazz, prompting Jordan to expand that sentiment: "Defense wins championships, without a doubt." When the Bulls "three-peated" in 1998, Jordan declared, "Defense wins championships; that's more evident than ever."

The importance of defense is so self-evident that the only debate appears to involve a matter of degree. Several years ago, a *Los Angeles Times* columnist declared, "Defense wins championships, especially in the NHL." A colleague at the *Contra Costa Times* begged to differ, writing, "Defense wins championships, especially in the NFL." A writer at the *Philadelphia Inquirer* specified further: "Defense wins championships, especially in the NFC." A Virginia Commonwealth hoops coach disagreed: "Defense wins championships, especially in basketball." It appears that *defence* wins championships, too, according to various Canadian hockey coaches, British football (soccer) managers, and Australian rugby personalities.

The sentiment has hardened from cliché into an article of sports law. But is it actually *true*?

We found that when it comes to winning a title, or winning in sports in general for that matter, offense and defense carry uncannily similar weight.

Among the 44 NFL Super Bowls, the better defensive team—measured by points allowed that season—has won 29 times. The better offensive team has won 24 times.* It's a slight edge for defense, but it's a pretty close call and not different from random chance. How many times has the Super Bowl champ been a top-five defensive team during the regular season? Twenty-eight. How many times was the Super Bowl champ ranked among the top five in offense? Twenty-seven. Nearly even.†

But we're talking about only 44 games, so let's broaden the sample size. There have been 407 NFL playoff games over the last 44 seasons. The better defensive teams have won 58 percent of them. The better offensive teams have won 62 percent of the time. (Sometimes, of course, the winning team is better both offensively and defensively, which explains why the total exceeds 100

* Note that that adds up to 53, which means that some teams are the better offensive *and* defensive team in the Super Bowl. In fact, 19 Super Bowls have featured a team superior on both sides of the ball. Those teams have won 14 of those games.

† It turns out the top-ranked defense during the regular season has won 15 Super Bowls, whereas the top-ranked offense has won only 8. Although this would seem to confer an advantage on defense, these two numbers are not statistically different. And given that the top-five defenses have won no more than the top-five offensive teams, it also means that offensive teams ranked 2–5 have won more Super Bowls than defensive teams similarly ranked, though again, these differences are not statistically significant.

percent.) That's a slight edge to the offense, but again, pretty even. Collectively, teams with a top-five defense have won 195 playoff games. Teams with a top-five offense have won 192 playoff games. In almost 10,000 regular season games, the better defensive team has won 66.5 percent of the time compared with 67.4 percent of the time for the better offensive team. That's a slight nod to the offense but a negligible difference.

But maybe the phrase "defense wins championships" means that defense is somehow more *necessary* than offense. Maybe a team can prevail with a middling offense, but not with a middling defense. As it turns out, that doesn't hold up, either. Three times the Super Bowl champion ranked in the bottom half of the league in defense; only twice did it rank in the bottom half in offense. The lowest-ranked defensive team to win a Super Bowl was the 2006 Indianapolis Colts, rated nineteenth that year. (They offset that by ranking third in offense.) The lowest-ranked offensive team to win the Lombardi Trophy? The 2000 Baltimore Ravens, who ranked . . . nineteenth in offense but first in defense. In the NFL, it seems, teams need *either* exceptional defense *or* exceptional offense to win a championship. But neither one is markedly more important than the other.

What happens when the best offenses line up against the best defenses—say, the 2006 Colts versus the 2000 Baltimore Ravens? It turns out that 27 Super Bowls have pitted a top-five offense against a top-five defense. The best offensive team won 13, and the best defensive team won 14. Another stalemate.

In the NBA, too, defense is no more a prerequisite for success than offense is. (Sorry, Michael.) Of the 64 NBA championships from 1947 to 2010, the league's best defensive teams during the regular season have won nine titles and the best offensive teams have won seven. That's pretty even. In the playoffs, the better defensive teams win 54.4 percent of the time and the better offensive teams win 54.8 percent of the time—almost dead even. Among 50,000 or so regular season games, the better defensive teams win no more often than the better offensive teams. We see the same results in the NHL. There's no greater concentration of Stanley

Cups, playoff wins, or regular season victories among the team playing the best defense/defence than among those playing the best offense.

Baseball is a bit tricky to analyze since "defense" includes pitching and is determined more by the guy on the mound than by the effort and dedication of the other eight players on the field. Still, there's not much evidence that defense is indispensable for winning a championship. Among the last 100 World Series winners, the superior defensive team has won 44 times and the superior offensive team has won 54 times. Among all postseason games, the better defensive teams have won 50.8 percent of the time versus 51.8 percent for the better offensive teams. That's remarkably even.

Okay, but does defense give an *underdog* more of a chance? Are upsets more likely to be sprung by defensive-minded teams? Sifting through the statistics, we found that the answer is no. We calculated that in the regular season, playoffs, and championships, underdog teams are no more likely to win if they are good defenders than if they are good scorers.

If defense is no more critical to winning than offense is, why does everyone from Little League coaches to broadcasters to Michael Jordan persist in extolling its importance? Well, no one needs to talk up the virtues of scoring. No one needs to create incentives for players to shoot more goals and make more jump shots or score runs and touchdowns. There's a reason why fans exhort "De-fense, De-fense!" not "Of-fense, Of-fense!" Offense is fun. Offense is glamorous. Defense? It's less glamorous, less glorified. Who gets the Nike shoe contracts and the other endorsements, the players who score or the defensive stoppers? And for all the grievances about today's "*SportsCenter* culture" that romanticizes dunks and home runs but ignores rebounds and effective pass rushing, the fact is that it's always been true. Which of the following sets of names is more recognizable? The top five touchdown leaders in NFL history: Jerry Rice, Emmitt Smith, La-Dainian Tomlinson, Randy Moss, and Terrell Owens? Or the top five interception leaders: Paul Krause, Emlen Tunnell, Rod Woodson, Dick Lane, and Ken Riley?

Players—especially younger players—need incentives to defend aggressively. The defense gets blamed if the team gives up a score or a basket but gets little praise if it does a good job—no matter how vital it might be to the narrative of the game. Think back to Michael Jordan. As long as he was on the floor, there was little concern that the Bulls would score points. Ultimately, Chicago's success was going to hinge on whether the team committed to rebounding and contesting shots and denying passing lanes. Jordan needed to encourage his teammates to commit to what's rightly called dirty work, the grit of tough defense. His frequent refrain of "defense wins championships" was a clever way of reinforcing the work—and work ethic—of his teammates. (And to Jordan's great credit, he played offense and defense with comparable excellence.)

But there may be something else at play, as well. Think back to loss aversion, the notion that we hate to lose more than we love to gain. On offense, athletes seek a gain. They're looking to score, to increase a lead or reduce a deficit, to change the numbers on the scoreboard. On defense, athletes are trying to prevent points, to preserve the score and keep it from changing. Perhaps if sports were structured differently, defense might be perceived differently. Imagine if every game started not at 0–0 but with a score of, say, 25–25, and teams could only have points *deducted* from that total. It stands to reason that the principles of loss aversion might kick in and inspire better defense the same way the prospect of a material *loss* of strokes inspires Tiger Woods to perform better on par putts than on birdie putts.

But the bottom line is this: Defense is no more important than offense. It's not defense that wins championships. In virtually every sport, you need *either* a stellar offense *or* a stellar defense, and having both is even better. Instead of coming with the "defense wins championships" cliché, a brutally honest coach might more aptly, if less inspirationally, say: "Defense is less sexy and no more essential than offense. But I urge it, anyway."

THE VALUE OF A BLOCKED SHOT

Why Dwight Howard's 232 blocked shots
are worth less than Tim Duncan's 149

His father was drafted by the San Francisco 49ers and played for the Calgary Stampeders of the Canadian Football League. His son played baseball in college. But John Huizinga had always gravitated to hoops. As a kid, he idolized Bill Russell and spent most of his teenage years playing pickup games in the gyms of San Diego. Sprouting to six feet, three inches tall, Huizinga played shooting guard for Pomona College in California. Later, he became a professor and eventually a dean at the University of Chicago's Booth School of Business. Yet he never kicked his basketball jones, managing a fantasy league team, the Dead Celtics, and devoting untold hours to watching NBA games.

Knowing how much Huizinga liked basketball, a colleague invited him to watch a top Chinese prospect work out in Chicago before the 2002 NBA draft.

"Yao Ming?" Huizinga asked.

"Yup," said the colleague, a Chicago statistics professor.

Huizinga was confused. He knew all about Yao, a seven-foot, six-inch center from China, projected as the top pick in the draft. But because of a thorny political situation and tense negotiations with the Chinese sports authorities, Huizinga knew that Yao's movements were shrouded in secrecy.

"How can we get in?" Huizinga asked his colleague. "I'm sure it's closed to the public."

"Easy," the professor explained. One of his MBA students, Erik Zhang, was a family friend of Yao's, tasked with running the workout. Zhang had wanted to reschedule his midterm so that he could help Yao impress the scouts. The professor told Zhang he could postpone the exam. As a gesture of thanks, Zhang agreed to sneak the professor and a friend into the workout.

After walking into a dingy gym in downtown Chicago, alongside Pat Riley, Huizinga watched as Yao displayed his dazzling skills before a small audience of NBA scouts, executives, and coaches. "It was a cool afternoon," says Huizinga. "I thought that was that." But then Zhang invited Huizinga along for dinner. Huizinga didn't know Chinese, but he knew basketball, and he knew all about negotiating. Zhang also realized that the Chinese authorities might be more receptive to dealing with a college professor than with a slick, in-your-face NBA agent. By the end of the meal, the group had suggested that Huizinga be Yao's representative.

So it was that John Huizinga, a University of Chicago professor by day, spent the better part of the last decade moonlighting as the agent for the Houston Rockets center, arguably the most popular basketball player on the planet. Huizinga traveled the world, negotiated more than $100 million in salary for his client, and haggled over the fine print on sneaker contracts. Says Huizinga: "It's also meant that I've watched more NBA basketball than you might think humanly possible."

In so doing, he began noticing something curious about blocked shots in basketball. Some of them, he believed, had much more value than others. A block of a breakaway layup? That's pretty valuable, since the opposing team is almost surely going to score. If the blocked shot is recovered by a teammate, who then starts his own fast break—a "Russell," to borrow the coinage of popular columnist Bill Simmons—well, that's even more valuable. After all, it not only prevents the opponent from scoring but leads to two points on the other end. Contrast this with a block of an awkward, off-balance leaner as the shot clock expires, or of a three-point attempt—that is, a shot much less likely to be successful. Or consider a shot swatted with bravado into the stands, enabling the opposing team to keep possession. Those blocks aren't nearly so valuable.

Huizinga calculated that if context were taken into account, fans and coaches might think differently about the NBA's top shot blockers. He teamed with Sandy Weil, a sports statistician, to examine the last seven seasons of NBA play-by-play data, focusing on the types of shots blocked (e.g., jumpers versus layups) as well as the outcomes from those blocks (e.g., tipping to a teammate versus swatting out of bounds). They estimated the block of an attempted layup or a "non jump shot" to be worth about 1.5 points to the team. Without the block, opponents score or draw fouls most of the time, resulting in 1.5 points on average. For jump shots, which go into the hoop with less frequency, the value of a block is only one point. And so on.

Huizinga and Weil also assigned a value to the outcome of each block. Blocking the shot back to the opponent was assigned one value. Blocking the ball out of bounds so that the opponent retained possession but had to inbound the ball was worth slightly more. Blocking the ball to a teammate was worth the most. Finally, they examined goaltending, the least valuable block for a team, as it not only guarantees two points to the opponent but occasionally results in a foul, leading to a three-point play.

Sure enough, Huizinga and Weil found that if you rank players on the *value* of their shot blocks, taking into account the types of blocks and the outcomes, it differs significantly from the NBA's list of the top shot blockers, which is simply numerical. As one glaring example, in 2008–2009, Orlando's abundantly talented, abundantly muscled center, Dwight Howard, blocked 232 shots, which factored heavily in his winning the NBA's defensive MVP award. Yet his accumulation of blocked shots was actually worth less, Huizinga and Weil calculated, than the 149 shots blocked by San Antonio's Tim

Duncan. How? It turned out that Howard often blocked shots into the stands, whereas Duncan often tipped the ball to a teammate. More important, Howard also committed goaltending violations more often than Duncan did. (In fact, Duncan, despite being a prolific shot blocker, hasn't goaltended in over three seasons.) Howard may have blocked 83 more shots than Duncan did, but they amounted to a value of only 0.53 points per block for the Magic. Duncan's average block was worth 1.12 points for the Spurs.

When the top shot blockers were reranked by the *value* of their blocks rather than by the sheer number, Duncan's status as a truly elite center was affirmed. Though he's never led the NBA in blocked shots, four times over the last decade he's posted the highest value-per-block totals. By comparison, Dwight Howard's best showing on a value basis is fifteenth. Milwaukee center Andrew Bogut, not known as a particularly fearsome shot blocker, delivers value. Mavericks big man Erick Dampier does not. And Huizinga's client, Yao Ming, falls squarely in the middle.

The following table shows the ten most valuable shot-blocking performances over the last eight NBA seasons and the ten least valuable performances on a value-per-block basis. Tim Duncan owns four of the ten most valuable performances; Dwight Howard owns three of the least valuable.

Of course, the total value of one's shot-blocking is the number of blocked shots times the value of the blocks. If all your blocks are Russells, the most valuable type of block, but you produce only a handful of them a year, your score will not be very high. Similarly, if you block a ton of shots but most of the blocks aren't that valuable, that isn't so useful to the team, either.

Although the study tried to account for as many factors as possible, it's not perfect, as Huizinga and Weil readily admit. There's no accounting, for instance, for the increased fouls a shot blocker can accumulate with overly aggressive play or the potential increases in offensive rebounds—when the team shooting recovers the ball—that occur when the shot blocker leaves his man to attempt a swat. The study also can't account for any intimidation factor: how

TEN MOST VALUABLE SHOT BLOCK
PERFORMANCES FROM 2002 TO 2009

PLAYER	SEASON	BLOCKS	VALUE PER BLOCK
Tim Duncan	**2008**	**149**	**1.12**
Andrew Bogut	2008	137	1.10
Rasho Nesterovic	2003	117	1.09
Zydrunas Ilgauskas	2006	136	1.08
Ben Wallace	2007	156	1.07
Ben Wallace	2008	120	1.07
Tim Duncan	**2005**	**171**	**1.06**
Chris Kaman	2006	110	1.05
Tim Duncan	**2006**	**162**	**1.05**
Tim Duncan	**2009**	**126**	**1.05**

TEN LEAST VALUABLE SHOT BLOCK
PERFORMANCES FROM 2002 TO 2009

PLAYER	SEASON	BLOCKS	VALUE PER BLOCK
Dwight Howard	**2008**	**232**	**0.53**
Erick Dampier	2003	165	0.60
Stromile Swift	2003	119	0.63
Erick Dampier	2004	146	0.66
Samuel Dalembert	2004	220	0.66
Dwight Howard	**2007**	**181**	**0.67**
Samuel Dalembert	2007	201	0.67
Joel Przybilla	2005	200	0.67
Shaquille O'Neal	2009	120	0.68
Dwight Howard	**2009**	**191**	**0.69**

many shots a player may deter with his mere presence, how many times he causes the shooter to change his trajectory and angle. Still, the research highlights that not all blocks are created equal. It's the value of an act, not the act itself, that ultimately matters.

That triggers a question: Why do we—and the NBA—*count* blocks rather than *value* blocks? The short answer: Counting is easy; measuring value is hard. We see this all the time in many facets of life and business. People count quantities (easy) rather than measure importance (hard) and as a result sometimes make faulty decisions. We award certificates for perfect attendance but seldom ask whether the winners learned more while in school. The associates promoted to partner in a law firm often are those who bill clients the most hours, but did they do the best work? We often care too much about how many stocks we own and not enough about the more relevant issue: the value of those stocks.

Sports, too, are filled with rankings based on simple numbers that don't always correspond to value. The interception of the nothing-to-lose Hail Mary pass on the play before halftime is worth far less than, say, Tracy Porter's game-sealing pick of Peyton Manning in the fourth quarter of Super Bowl XLIV. But the stats don't distinguish it as such. The value of an empty-net goal in hockey—when a losing team removes its goalie to have an extra skater on offense—isn't nearly as important as a decisive overtime goal. Yet the stat sheet doesn't make a distinction between them. Savvy general managers can recognize (and exploit) this kind of information, acquiring and unloading talent accordingly, much the way investment managers look for undervalued securities to buy and overvalued ones to sell.

After the 2009–2010 NBA season, Dwight Howard was named the league's best defensive player for the second year in a row. The vote was a landslide: Howard received 110 out of 122 first place votes. Howard led the league in both rebounds (13.2) and blocked shots (2.78) per game, the first player to lead in both categories for two consecutive years since the NBA started tracking blocked shots in the 1973–1974 season. But the voting surely would have been closer if his blocks had been valued and not simply tallied. Howard's coach, Stan Van Gundy, observed: "I think people see the blocked shots and they see the rebounding, but I don't think unless you're a really astute observer that they see the other things he does for us defensively."

Other astute observers, though—basing their judgment on value, not raw quantity—might reach a less flattering conclusion.

ROUNDING FIRST

Why .299 hitters are so much more rare
(and maybe more valuable) than .300 hitters

Whether we're buying batteries at Walmart, a fast-food value meal, or even a house, odds are good that the price ends in a nine. We're numb to seeing $1.99 bottles of Coke, $24,999 cars, and even $999,999 McMansions on cul-de-sacs. In the case of gasoline, the price even extends to nine-*tenths* of a cent, say, 2.99^9 for a gallon of unleaded. This entire concept, of course, is silly. Purchase one gallon of 2.99^9 gas and it will cost you $3.00. It takes ten gallons before you realize any savings—and it's a mere penny at that—over gas priced at an even three bucks.

The difference between a price ending in a nine and one ending in a whole number is virtually meaningless, accounting for a negligible fraction of the purchase price, but test after consumer test reveals that there is great psychological value in setting a price point just below a round number. Even among sophisticated consumers who recognize the absurdity of it all, paying $9.99 is still somehow more palatable than paying $10.00. (Factor in sales tax and you're paying over $10 in both cases, which makes it more absurd.) Round numbers are powerful motivators—whether it's to hit them or avoid them—in all sorts of contexts.

Devin Pope and Uri Simonsohn, then a pair of Wharton professors, examined the prices of millions of used cars and found something that was at once peculiar and predictable: When the mileage on the vehicles eclipsed 100,000, the value dropped drastically. A car with 99,500 miles might have sold for $5,000, but once the odometer of that car—identical year, model, and condition—rolled over 500 more times and posted 100,000 miles, the value fell off a cliff. Why? Because customers for a used car set a benchmark of 100,000 miles, and woe unto the seller whose jalopy eclipsed that number.

When the two economists looked at the market for jewelry, they saw that pieces are sold as full karats and half karats but almost never as 0.9 karats. Why? Because shoppers have set a goal of a round number—"I want to buy her at least a two-karat ring"— and don't want to come up a little bit short. To do so would make them feel they had shortchanged the intended recipient.

In looking at human behavior, Pope and Simonsohn found that we're slaves to round numbers. Every year more than a million high school students take the SAT, aiming for a round-numbered score as a performance goal. How do we know this? Until 2005, the SATs were scored between 400 and 1600 in intervals of 10. When students posted a score ending in a 90 (1090, 1190, 1290, etc.), they were 20 percent more likely to retake the test compared with students whose score ended in a round number (1100, 1200, 1300). The difference in the scores might be as small as a single question, and according to Pope and Simonsohn, those ten points do not disproportionately change an applicant's chance of admission. Still, it meant everything to many teenagers (perhaps because they figured schools would have round score cutoffs). The most noticeable difference in students who decided to retake the test? It was between those scoring 990 and those scoring 1000.

Some of the most arresting results came when the researchers considered the behavior of Major League Baseball players. Baseball, of course, is flush with "round number targets." Pitchers strive for 20-win seasons. Ambitious managers challenge their teams to win 100 games. Hitters try like hell to avoid the notorious "Mendoza Line" of a .200 batting average. But no benchmark is more sacred than hitting .300 in a season. It's the line of demarcation between all-stars and also-rans. It's often the first statistic cited in making a case for or against a position player in arbitration. Not surprisingly, it has huge financial value. By our calculations, the difference between two otherwise comparable players, one hitting .299 and the other .300, can be as high as 2 percent of salary, or, given the average Major League salary, $130,000. (Note that though the average MLB salary is $3.4 million, it's closer to $6.5 million for players batting in the .300 range.) All for .001 of a batter's average, one extra hit in 1,000 at-bats.

Given the stakes, hitting .300 is, not surprisingly, a goal of paramount importance among players. How do we know this? Pope and Simonsohn looked at hitters batting .299 on the final day of each season from 1975 to 2009. One hit and the players could vault above the .300 mark. With a walk, however, they wouldn't be credited with an at-bat or a hit, so their averages wouldn't budge. What did these .299 hitters do? They swung away—wildly.

FREQUENCY OF WALKS
DURING LAST AT-BAT OF SEASON

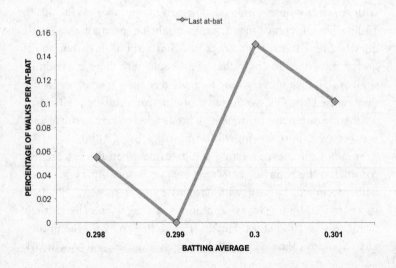

We looked at the same numbers, and here's what we found. Players hitting .300 walked 14.5 percent of the time and players hitting .298 walked 5.8 percent of the time, but in their final plate appearance of the season, players hitting .299 have *never* walked. In the last quarter century, *no player hitting .299 has ever drawn a base on balls in his final plate appearance of the season.*

The following chart highlights these numbers. Note that it spikes like the EKG of a patient in cardiac arrest.

If we look at the likelihood of a walk for hitters just below .300 versus just above .300 before the last game of the season—or even during the last game but before the last at-bat—we don't see any stark differences. But for that last at-bat, when they're desperate to reach that .300 mark, they refuse to take a base on balls, swinging away to get that final hit that will put them over the line. The following chart highlights this, even indicating that before the last game, .299 hitters actually walk slightly *more* than .301 hitters.

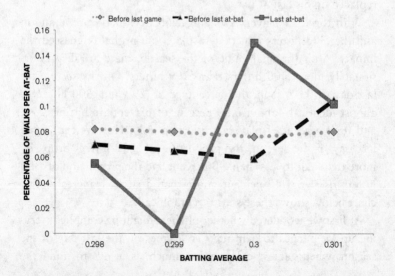

**FREQUENCY OF WALKS
DURING LAST GAME OF SEASON**

What's more surprising is that when these .299 hitters swing away, they are remarkably successful. According to Pope and Simohnson, in that final at-bat of the season, .299 hitters have hit almost .430. In comparison, in their final at-bat, players hitting .300 have hit only .230. (Why, you might ask, don't *all* batters employ the same strategy of swinging wildly, given the success of .299 hitters? Does this not indict their approach the rest of the season? We think not. For one thing, these batters never walk, so their on-base percentages are markedly lower than those of more conservative hitters. Also, if *every* batter swung away liberally throughout the season, pitchers probably would adjust accordingly and change their strategy to throw nothing but unhittable junk.)

Another way to achieve a season-ending average of .300 is to hit the goal and then preserve it. Sure enough, players hitting .300 on the season's last day are much more likely to take the day off than are players hitting .299. Even when .300 hitters do play, in their final at-bat they are substituted for by a pinch hitter more than 34 percent of the time. In other words, more than a third of the time, a player hitting .300—an earmark of greatness—will relinquish his last at-bat to a pinch hitter. (Hey, at least his average can't go down.) By contrast, a .299 hitter almost never gets replaced on his last at-bat.

With the .299 players swinging with devil-may-care abandon and the .300 hitters reluctant to play, you probably guessed the impact: After the final game of the season, there are disproportionately more .300 hitters than .299 hitters. On the *second*-to-last day of the season, the percentage of .299 and .300 hitters is almost identical—about 0.80 percent of players are hitting .299, and 0.79 percent of players are hitting .300. However, *after* the last day of the season, the proportion of .299 hitters drops by more than half to less than .40 percent and the proportion of .300 hitters rises to 1.40 percent, more than a twofold increase. The chart below shows these statistics graphically.

At first we wondered whether pitchers might be complicit, serving up fat pitches to help their colleagues on the last day of the season, when games seldom mean much. It brings to mind the

DISTRIBUTION OF MLB PLAYERS' BATTING AVERAGE BEFORE, DURING, AND AFTER LAST AT-BAT OF SEASON

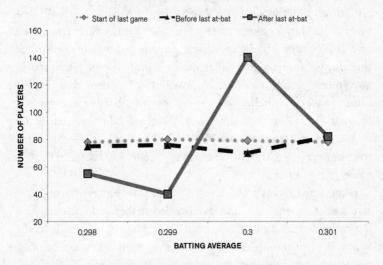

batting race of 1910, one of the great controversies in baseball history. That year, the Chalmers Auto Company promised a car to the player who won the batting crown. Ty Cobb was leading by nine points heading into the final game of the season, and much as some players still do today, Cobb took the day off to protect his lead. Cobb was a contemptible figure, a virulent racist disliked by most of his fellow players, including his own teammates. Cleveland infielder Nap Lajoie was second in the batting race and far more popular than Cobb. On the season's final day, playing against the St. Louis Browns, Lajoie went eight for eight in a doubleheader, including seven bunt hits that "dropped" in front of a third baseman who had been positioned by his manager to play in short left field. When Lajoie reached safely on one bunt that was ruled a sacrifice, the Browns brass offered the official scorer inducements to reconsider. (He declined, and the executives who tried to bribe him were effectively kicked out of baseball for life.) Despite the generosity/complicity of the opponents on the season's final day, Lajoie lost out to Cobb, .385 to .384.

We found no evidence of similar collusion today between pitchers and .299 batters on the final day of the season. Even if the pitchers are aware that the hitter they're facing is just below the .300 threshold, there's no indication they're tossing meatballs destined to be hit. Looking at the Pitch f/x data, which tracks not only the location but the speed, movement, and type of every pitch thrown, we found that neither the location, type of pitch, speed, movement, or any measureable attribute of pitches was reliably different when a pitcher faces a batter with a batting average just below the .300 mark at the end of the season. Pitchers either are unaware that batters are just shy of .300 or don't care—they pitch the same way to a .299 hitter as they do to a .300 batter in the last game of the season.

Data, however, tell us what isn't the same: Batters hitting .299 swing more liberally, taking fewer called strikes and balls but having more swings and misses, even when facing three balls in the count. In their last at-bat they did everything possible *not* to draw a walk. As a result, they got more hits. In comparison, .300 batters drew many walks and did not swing on three-ball counts.

■ ■ ■

Of course, the .300 mark isn't the only round number players strive to attain. The century mark for runs batted in (RBIs) is another coveted goal, and we see the same pattern, with many players ending the season with 100 RBIs and few ending with 99. Again, the differences are generated mostly from the last game and even the last at-bat of the season. First, players stay in the game long enough to hit their hundredth RBI, skewing the numbers. Remove that last at-bat and the prevalence of players with 99 RBIs is equal to that of those with 100.

The same pattern emerges for players with 19 or 29 (or 39 or 49) home runs. Players do everything possible to move up to the next round number, and so on their last at-bat, they "swing for the fences." We see a disproportionate number of players hitting 20 or 30 (or 40 or 50) home runs compared with 19 or 29 (or 39 or 49) home runs, often thanks to that last at-bat.

We see the same thing with pitchers trying to reach 20 wins for the season. Managers will even use their starting pitchers in relief toward the end of the season to boost their win totals. As with .300 versus .299 hitters, year to year we see more 20-game winners than 19-game winners.

Like achieving a .300 batting average, the difference between having 99 versus 100 RBIs, or 29 versus 30 home runs, or 19 versus 20 wins as a pitcher is worth real money in terms of future salary. Teams, owners, and general managers (GMs) clearly value the higher round numbers. In this respect, players may be acting economically rationally, responding to the incentives provided by teams to do everything they can on their last at-bat to reach those numbers.

We also noticed something else. The numbers above and those from the study looked back only to 1975, a period during which free agency was in force.* We took a look at the data going way back before free agency (prior to the early 1970s). We found that the number of players hitting round numbers exactly, relative to those just missing them, diminished significantly before the free agent era, another clue that players are responding to the financial lure of round numbers.

The puzzle is *why* the Republic of Sports values round numbers so much. One might even contend they should value round numbers *less* because they are being gamed by the players. Is the extra salary paid for a .300 hitter really justified when it is determined largely by a single at-bat on the last play in what was probably a meaningless game at the end of the season or when it was attained by sitting out the last game to ensure that the player's average didn't go down? The difference between a .300 and a .299 hitter is negligible—one misdirected ground ball, one blooper into short center field, one random bounce, one generous judgment by an official scorer over the course of a season. We would argue that .300 and .301 hitters

---

* Free agency allows players to negotiate and sign with other teams once their contracts expire. Previously, teams could invoke "reserve clauses" that allowed them to repeatedly renew a player's contract for one or more years and did not allow the player to terminate it.

are overvalued and .298 and .299 hitters are undervalued. A hedge fund manager would spot this as an "arbitrage" opportunity and unload the overvalued asset and buy the undervalued one. A savvy GM might consider doing the same thing: trading the .300 hitter for a player who hit just under .300, saving many thousands without affecting the hitting performance of his lineup.

■ ■ ■

Want an example of a player who has benefited greatly from round numbers? In 2003, remarkably, Bobby Abreu, then of the lowly Phillies, entered the last game of the season hitting .299 with 20 home runs and 99 RBIs, just shy of two prized benchmarks. The Phillies, long since eliminated from the playoffs, faced the Atlanta Braves, a team that had clinched the NL East title several weeks before. In the first inning, Abreu hit a groundout to first base that scored a runner on third, giving him 100 RBIs for the season. However, the groundout lowered his average to .2986 (172 hits in 576 at-bats). Coming up to bat again in the bottom of the third inning, he singled to center, driving in another run, giving him a .300 average for the season. He was taken out of the game before he could bat again.

A year later, in 2004, Abreu again entered the last game of the season hitting .299, and once again the Phillies were out of playoff contention. In his first at-bat he doubled, giving him a .300 average (172/573) for the season. However, he also had 29 home runs, so he continued playing to try to hit his thirtieth. After his hit in the first inning, he had a cushion—he could make an out and still keep his average above .300. (If he failed to get a hit, his average would drop to .2997—172 hits in 574 at-bats—and still round up to .300.) In the bottom of the third inning, Abreu went deep on an eight-pitch at-bat, hitting his thirtieth home run *and* getting a .301 average for the season. And what did he do? You guessed it. He left the game before he could bat again.

In his dozen full Major League seasons, Bobby Abreu has had *five* seasons finishing with exactly 20 or 30 home runs, *seven* seasons finishing with 100 to 105 RBIs, and *no* seasons with 95 to 99

RBIs. Put it this way: Surely, breaking these thresholds didn't hurt his contract negotiations. In 2002, he was awarded a five-year, $64 million contract with a $3 million signing bonus. With the previous contract he was making $14.2 million over three years.

Athletes and fans also care deeply about the milestones of career statistics. No elite baseball player wants to end his career with 999 RBIs or 299 pitching victories. (Not for nothing is the Bernie Mac movie called *Mr. 3000,* not *Mr. 2999.*) And no career stat seems to get more attention than home runs. Whenever a player nears a benchmark in home run totals, the milestone becomes a millstone. In the summer of 2010, Alex Rodriguez, the Yankees' controversial slugger, hit his 599th home run in his 8,641st at-bat, a clip of one dinger every 14.4 plate appearances. However, it took him 47 at-bats—including 17 straight hitless ones—to finally reach his 600th home run on August 4, 2010. (It also took him 29 at-bats to go from 499 to 500.) Yet after hitting his 600th home run, A-Rod went on a five-game hitting streak; within the next ten days, he'd hit home runs 601, 602, 603, and 604. Why was going from 599 to 600 more important than going from 598 to 599 or 600 to 601?

As you might have guessed, it's not just baseball that reveres round numbers. In the NFL, rushing for 1,000 yards is a benchmark every running back strives to achieve. A 1,000-yard rusher is perceived to be worth more than his 990-yard counterpart, and so running backs entering the last game just under 1,000 yards naturally get the ball more often than normal (18.3 versus 14.7 carries on average) and rush for more yards than normal (78 versus 62.5). Players with just over 1,000 yards run the ball in their last game about the same as they always do—67.2 yards on 15.5 carries—which is less than the numbers for their counterparts who started the day on the short side of 1,000 yards.

■ ■ ■

Of course, part of what we see in professional sports is driven by incentives. Although most sports and teams frown on bonus clauses—in which players are rewarded for hitting statistical benchmarks that might put individual interests before team interests—players do stand to gain in future contracts if they break certain thresholds. The question is, what's so special about those benchmarks? Is achieving 1,000 yards rushing really that much more valuable than getting 999? The lure of round numbers creates artificial barriers and causes us to overemphasize and overvalue them and, more to the point for general managers, undervalue those who barely miss the mark.

The same is true, of course, of areas outside sports. In the financial markets, we see perhaps the most extreme example of targeting numbers. Every fiscal quarter corporations announce their earnings numbers, which are compared to target earnings set by analysts' consensus forecasts on Wall Street. Beat your earnings forecast by a penny or two a share and your stock price will rise. Miss it and see that price plummet. But of course the target is just an estimate, with plenty of chance for error. Beating it or missing it by a little is likely to be the result of good or bad luck—in other words, randomness. Surely, failing to hit a quarterly target by a few pennies shouldn't matter in the long run, should it?

Tell that to corporate executives. Among the tens of thousands of earnings announcements and earnings estimates each quarter, very few firms miss by a penny. Just like aspiring .300 hitters, a huge number of firms each year meet their targets *exactly* or exceed them by a penny. How do they do this? Well, accounting rules offer some discretion for the way corporations deduct expenses, report income, depreciate assets, and so on, and all these things can be used to alter the bottom line slightly. If its earnings are going to fall a bit short, a corporation may take a tax deduction this quarter rather than next or defer the expense on new equipment until next year to bump up earnings this year. The following graph, from an academic paper by Richard Frankel and Yan Sun from Washington University in St. Louis and William Mayew of Duke University, highlights this phenomenon. Plotting the frequency of corporate earnings "surprises"—the difference between the actual earnings numbers and the consensus forecast set by analysts—shows that a disproportionate number

FREQUENCY OF FIRM
QUARTERLY EARNINGS SURPRISES

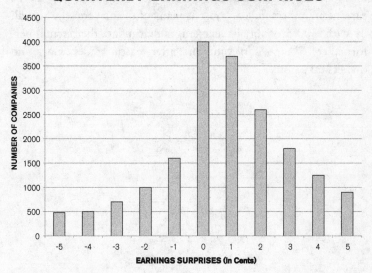

of firms exactly meet their target, having zero earnings surprises. Another huge percentage of firms beat their targets by exactly one penny. In contrast, very few firms miss by a penny. From a statistical standpoint this is extraordinary. Randomness suggests that as many firms would miss by one cent as would meet or beat the target by one cent.

If the graph looks eerily familiar, it should—it resembles the number of .300 hitters plotted against the paucity of those hitting .299. Corporations, like professional athletes, continually work to "manage" (or game) their performance numbers. Athletes do this each year before the season ends, corporations each quarter before they're evaluated by investors and analysts.

Early on, investors in the financial market, just like sports GMs, seemed to play along. Miss your target by a cent and your stock price plummets. Beat your target by a penny and your stock price rises. However, investors have caught on. Today, just meeting an earnings target isn't enough. The stock price will drop. Why? Because investors have figured out that just meeting the target probably means the firm did everything it could to make its earnings look good. Translation: The news is not quite as rosy as the earnings numbers indicate, just as many of the players who squeak by with .300 probably have averages that inflate their actual performance. Come salary time and after-season trades, GMs and owners take note.

THANKS, MR. ROONEY

Why black NFL coaches are doing worse than ever—and why this is a good thing

Even without the benefit of hindsight, Tony Dungy seemed to be the perfect representation of what NFL teams look for in a head coach. He carried himself with a quiet but towering dignity, at once firm and flexible, stern and compassionate, fully committed to his job, his family, and his faith. His players revered him. His assistants aspired to *be* him. Even the doctrinaire members of the media spoke of him in glowing terms. Mel Blount, the Pittsburgh Steelers' Hall of Fame cornerback, played with Dungy in the late 1970s. "Even then you knew it," says Blount. "Tony was born to be a head coach in the NFL."

After a long stint—too long, many thought—as an NFL assistant, Dungy, an African American, finally got his chance. The Tampa Bay Buccaneers hired him as head coach in 1996. Although Dungy acquitted himself well, he couldn't alchemize his passion and professionalism into victories—at least not enough of them. In six seasons, Dungy's teams won more than half their games and he took the Bucs to the playoffs four times, but the teams struggled once they got there. Two days after a 31–9 defeat to the Philadelphia Eagles in the 2001 postseason, Dungy was relieved of his coaching duties—a decision that seemed validated when his successor, Jon Gruden, coached the team to a Super Bowl win the next season.

At the time, Dungy's firing left the NFL with just two African-American head coaches, roughly 6 percent. On its face, it was a dismal record, especially when you considered that African Americans made up nearly three-quarters of the league's players. And this wasn't an "off year." In 1990 and 1991 there was just one African-American head coach in the NFL. From 1992 to 1995 there were two. There were three between 1996 and 1999, and there were two in 2002. This struck many as wrong, but statistics alone weren't enough to show bias. One could just as easily claim that the disproportionately small pool of white players was, statistically anyway, more anomalous. It wasn't unlike the English Premier League in soccer, where 75 percent of the coaches are British but the majority of the players come from outside England.

Yet Johnnie Cochran Jr.—the controversial lawyer remembered best for his glove-doesn't-fit defense of O. J. Simpson—joined forces with another activist attorney, Cyrus Mehri, and decided to challenge the NFL's hiring practices. At the time, Cochran and Mehri had been working on a case targeting what they saw as biased employment practices at Coca-Cola. In the course of the Coke case, they had crossed paths with Janice Madden, a sociologist at the University of Pennsylvania specializing in labor economics. Madden was in Atlanta, working on the same case, using a statistical model to demonstrate that women were not, as the company alleged, inferior salespeople. A thought occurred to Cochran and Mehri: Maybe Madden could initiate a similar study with respect to NFL coaches.

Although Madden shares a surname with former NFL coach, popular NFL announcer, and video game impresario John Madden, the football similarities ended there. She was not much of a fan. Her husband was a Philadelphia Eagles season ticket holder, but she preferred to spend her Sundays at home. Still, she made Cochran and Mehri an offer: "If you can put the data together for me, I'll do this pro bono." They did, and she did.

Madden found that between 1990 and 2002, the African-American coaches in the NFL were statistically far *more* successful than the white coaches, averaging nine-plus wins a season versus eight for their white counterparts. Sixty-nine percent of the time, the black coaches took their teams to the playoffs, versus only 39 percent for the others. In their *first* season on the job, black coaches took their teams to the postseason 71 percent of the time; rookie white coaches did so just 23 percent of the time. Clearly, black coaches had to be exceptional to win a job in the first place.

Perhaps, one could argue, black coaches ended up being offered jobs by the better teams: the franchises that could afford to pursue talent more aggressively. Madden reran her study, controlling for team quality. African-American coaches still clearly outperformed their colleagues. If this wasn't a smoking gun, to Madden's thinking, it surely carried the strong whiff of bias. If African-American football coaches were being hired fairly, shouldn't they be performing comparably to white coaches? The fact that the win-loss records of African-American coaches were substantially better suggested that the bar was being set much higher for them.

When Madden went public with her findings, she was blindsided by the criticism. The NFL made the argument that Madden's sample size—in many seasons there were just two African-American coaches—was too small to be statistically significant. Whose fault was that? Madden wondered. At the national conference for sports lawyers, an NFL executive dismissed Madden's work, suggesting that she could have run the numbers for "coaches named Mike" and for "coaches not named Mike" and come up with similar results. (Curious, Madden ran the numbers and found that this wasn't the case.)

Still, due in no small part to the work of a female sociologist whose football knowledge was admittedly modest, the NFL changed its ways. In 2003, the league implemented the so-called Rooney Rule, named for Dan Rooney, the progressive Steelers owner who chaired the committee looking into the issue. The rule decreed that teams interview at least one minority applicant to fill head-coaching vacancies. Otherwise, the franchise would face a stiff fine.

In 2003, the NFL levied a $200,000 fine against the Detroit Lions when the team hired Steve Mariucci without interviewing any other candidates, black or white. (Mariucci went 15–28 and was fired in his third season.) The league achieved its aim. By 2005, there were six African-American coaches in the NFL, including Dungy, who had been hired by the Indianapolis Colts.

And how has this new brigade of black coaches done? Worse than their predecessors. Much worse, in fact. From 2003 to the present, African-American coaches have averaged the same number of wins each season—eight—as white coaches. They are now slightly *less* likely to lead their teams to the playoffs. Their rookie seasons are particularly shaky: They lose slightly more games than white coaches do in the first season. In 2008, for instance, Marvin Lewis coached the Cincinnati Bengals to a 4–11–1 record, which was only slightly better than the job Romeo Crennel did a few hours' drive away in Cleveland, where the Browns stumbled through a 4–12 season. Lewis and Crennel still fared better than yet another African-American coach in the Midwest, Herman Edwards, who oversaw a misbegotten Kansas City Chiefs team that went 2–14.

It's worth pointing out that Crennel and Edwards were fired. The Bengals stuck with Lewis, and he promptly won NFL Coach of the Year honors in 2009, guiding Cincinnati to an unexpected 10–6 season. But as black coaches lose more games, Madden and other supporters nod with satisfaction. This "drop-off" is the ultimate validation of the Rooney Rule, an indication that black coaches are being held to the same standards as their white counterparts. "If African-American coaches don't fail, it means that those with equal talents to the failing white coaches are not even getting the chance to be a coach," Madden explains. "Seeing African-American coaches fail means that they, like white coaches, no longer have to be superstars to get coaching jobs."

The Tampa Bay franchise that fired Dungy and replaced him with Jon Gruden? When the team let go of Gruden in 2009, management replaced him with Raheem Morris, then a 32-year-old African American who was the team's defensive backs coach and had never before been a head coach on any level. Although no one admitted it, Morris was precisely the type of candidate unlikely to have been taken seriously before the Rooney Rule. In Morris's first season, the Bucs went 3–13.

Amid the surge in losing, there have been triumphs. In Super Bowl XLI, Dungy coached against Lovie Smith of the Chicago Bears, the second time two black coaches in a major American professional sport had faced each other for a championship and a first for the NFL. Dungy would finally get his Super Bowl ring. Two years later, the Pittsburgh Steelers, orchestrators of the Rooney Rule, prevailed in Super Bowl XLIII—an example of a good deed going unpunished. The team's coach was a Dungy disciple, Mike Tomlin. Yes, he is "a coach named Mike." He also is an African American.

COMFORTS OF HOME

How do conventional explanations for the
home field advantage stack up?

It was one of those games that get lost in the folds of an NBA season schedule, a thoroughly forgettable midweek, midseason clash
between the Portland Trail Blazers and the San Antonio Spurs. In
the late afternoon of February 25, 2009, the bus carrying most of
the Portland players arrived in the loading dock of San Antonio's
AT&T Center, surely the only arena in the league that carries
the faint odor of a rodeo. The Blazers had played—and lost—in
Houston the previous night, and as the players slogged through
the catacombs of the arena, headphones wrapped around their
ears, they wore the vacant, exhausted looks of employees grinding
through a business trip.

Which, you could contend, they were. At the end of the day—an
end that couldn't come soon enough for them—they were just another pack of salaried employees engaged in business, a thousand
miles from home, sleeping in strange beds, eating bland room service food thanks to a generous per diem, staring at a string of unfa-
miliar faces. It's a truism of business travel: Even when you go first
class—and Lord knows the Blazers, a team owned by billionaire
Microsoft cofounder Paul Allen, go first class—it still isn't home.

The Blazers and the Spurs had nearly identical records at the time, and owing to injuries, the Spurs were missing two of their three stars, Tim Duncan and Manu Ginobili, leaving only quicksilver point guard Tony Parker and a slew of role players. But it hardly mattered. The Las Vegas line predicted that the Spurs would win by as many as nine points. The Spurs' television broadcast began with an upbeat intro: "No Tim? No Manu? No problem!" Sitting casually on the scorer's table before the game, a Portland assistant coach cackled and confided matter-of-factly, "Ain't no way we're winning this motherf——."

Hearing those pronouncements, so favorable for San Antonio and so unfavorable for Portland, one could be forgiven for wondering whether the rules of basketball were somehow different on the road, whether the height of the goals or the number of points awarded for a made basket changed once a team left home. The Blazers could have been (should have been?) brimming with confidence and optimism, relishing the chance to beat the Spurs, a team that had an almost identical win-loss record but was playing shorthanded. Instead, the Blazers projected the same kind of defeatism and anticipated doom that a conservative Republican political candidate might feel campaigning in San Francisco.

Then again, maybe the collective fear and loathing was well placed. For all the conventional sports wisdom that can be disproved, deconstructed, or, at the very least, called into question, home team advantage is no myth. Indisputably, it exists. And it exists with remarkable consistency. Across all sports and all levels, going back decades, from Japanese baseball to Brazilian soccer to college basketball, the majority of the time the team hosting a game will win.

FIRST, THE FACTS

Consider the following table, which documents the home field advantage across 19 different sports leagues covering more than 40 countries, going back as far as we could (ordered from highest to lowest advantage by sport).

HOME FIELD ADVANTAGE

	% OF GAMES WON BY HOME TEAM	YEARS	% WON BY HOME TEAM LAST 10 YEARS
International Soccer			
MLS (U.S.)	69.1%	2002–2009	69.1%
Serie A (Italy)	67.0%	1993–2009	65.7%
Central America	65.2%	2001–2009	65.2%
La Liga (Spain)	65.0%	1993–2009	64.3%
South America	63.6%	2003–2009	63.6%
English Premier (U.K.)	63.1%	1993–2009	63.0%
Europe	61.9%	2000–2009	61.9%
Asia/Africa	60.0%	2005–2009	60.0%
Basketball			
NCAA (collegiate)	69.1%	1947–2009	68.8%
NBA	62.7%	1946–2009	60.5%
WNBA	61.7%	2003–2009	61.7%
Cricket			
International cricket	60.1%	1877–2009	57.4%
Rugby			
International rugby	58.0%	1871–2009	56.9%
Hockey			
NHL	59.0%	1917–2009	55.7%
Football			
NCAA (collegiate)	64.1%	1869–2009	63.0%
NFL	57.6%	1966–2009	57.3%
Arena football	56.0%	1987–2008	56.5%
Baseball			
MLB	54.1%	1903–2009	53.9%
Nippon League (Japan)	53.3%	1998–2009	53.6%

The advantage exists in all sports to varying degrees. The home team wins 54 percent of the time in Major League Baseball, nearly 63 percent of the time in the NBA, nearly 58 percent in the NFL, and 59 percent in the NHL. College basketball teams boast a whopping 69 percent home team success rate, and NCAA football confers a nearly as impressive 64 percent home team advantage. Across 43 professional soccer leagues in 24 different countries spanning Europe, South America, Asia, Africa, Australia, and the United States* (covering more than 66,000 games), the home field advantage in soccer worldwide is 62.4 percent. For nearly every rugby match in more than 125 countries dating back to as early as 1871, the home field advantage is 58 percent. For international cricket dating back to as early as 1877, covering matches from ten countries, the home winning percentage is 60 percent.

As radically as sports have changed through the years—the introduction of a three-point line in basketball, the addition of a designated hitter in MLB, the ever-escalating size of football players wearing helmets made of material other than leather—the home field advantage is almost eerily constant through time. In more than 100 baseball seasons, not once have the road teams collectively won more games than the home teams. The lowest success rate home teams have ever experienced in a baseball season was 50.7 percent in 1923; the highest was 58.1 percent in 1931.

In every season of play in the NBA, the NHL, and international soccer leagues, collectively the home teams have won more games than the road teams. In 43 of the 44 NFL seasons, home teams won at least 50.8 percent of their games. (In only one anomalous year, 1968, did home teams win less than half the games, mostly because there were five ties that season.) In 140 seasons of college football, there has never been a year when home teams have failed to win more games than road teams. The size of the advantage is remarkably stable in each sport, too: The home team's success rate is almost exactly the same in the last decade as it was 50 or even 100 years ago.

Another curious feature of the home field advantage: It is essentially the same within any sport, no matter where the games

---

* They include every league in Uruguay, Australia, Paraguay, Scotland, Japan, South Africa, England, Argentina, the Netherlands, Spain, Greece, Germany, Chile, Mexico, Italy, Honduras, Russia, Costa Rica, Brazil, France, El Salvador, Peru, Venezuela, and the United States.

are played. The Nippon Professional Baseball League in Japan has a home field advantage almost identical to that of Major League Baseball in the United States. The home field advantage in Arena football is virtually the same as it is in the NFL. The home winning percentage in the NBA is a virtual mirror image of that in the WNBA. In professional soccer, the sport with the largest home field advantage, hosting teams in three of Europe's most popular leagues—the English Premier, the Spanish La Liga, and the Italian Serie A—win about 65 percent of the time. Looking at 40 other soccer leagues in 24 different countries, the home field advantage hovers around 63 percent. All these statistics pertain only to the regular season in each sport, but the numbers are almost exactly the same for the playoffs.*

Some sports are even set up to give the home team an inherent advantage. In baseball, the home team bats last, so it always comes to bat knowing precisely how many runs it must score to win the game and can devise a strategy accordingly. (But notice that baseball has the *lowest* home winning percentage of the major sports, so that can't be a primary explanation.)

Another way to look at the home field advantage is to identify *how many* teams win more games at home than on the road. In the NBA, an astounding 98.6 percent of teams fare better at home than on the road. That means that in most seasons *all* NBA teams have better home than road records. At the time of the Blazers-Spurs game, Portland was 23–5 at home and 12–16 on

---

* For the playoffs an issue that has to be taken into account is that teams are typically seeded so that better teams get to play at home and worse teams are on the road more often. For this reason alone we would expect the home team to win more often in the playoffs. That is, home teams would win more of their fair share of games no matter where they played. Therefore, to compute the home field advantage in the playoffs accurately, we have to adjust for the quality of teams. Specifically, if Team A hosts Team B and Team A is a much better team, we first calculate how often you'd expect Team A to win if it played on a neutral field and compare that to how often Team A actually beats Team B when playing at home. The results? If you adjust for team quality, the home field advantage is almost *exactly* the same during the playoffs as it is during the regular season: For MLB it is 54 percent, for the NBA it is 61 percent, for the NFL it is 57 percent, and for the NHL it is 57 percent. These numbers are, once again, remarkably consistent.

the road, an 82 percent winning percentage at home versus 43 percent on the road. During that 2008–2009 season, only one team in the entire league would fare better on the road than at home, the lowly Minnesota Timberwolves, who were comparably awful regardless of venue: 11–30 at home and 13–28 away from Minneapolis. Most teams each year are similar to the Chicago Bulls of that season, who went 28–13 at home and the exact inverse, 13–28, on the road. Similarly, in hockey and soccer, more than 90 percent of the teams fare better at home than on the road. Even in the NFL and MLB, the leagues with the lowest home winning percentages, more than three-quarters of teams do better at home.

It's little wonder that leagues reward the best teams in the regular season with "home field advantage" for the playoffs—it's a hell of an incentive to win those dreary midseason games. And no wonder those playoff teams talk openly of aiming to achieve a "split on the road," essentially conceding the unlikelihood of winning multiple games away from home. When teams are down by a small margin in the final seconds of a game, there is even an adage, "Play for the tie at home and the win on the road." Think about this for a moment: Teams in an identical situation will strategize differently solely on the basis of *where* the game is being played.

Before considering the causes of the home field advantage, keep this premise in mind: There is considerable economic incentive for home teams to win as often as possible. When the home team wins, the consumers—that is, the ticket-buying fans—leave happy. The better the home team plays, the more likely fans are to buy tickets and hats and T-shirts, renew their luxury suite leases, and drink beer, overpriced and watered down as it might be. The better the home team plays, the more likely businesses and corporations are to buy sponsorships and the more likely local television networks are to bid for rights fees. A lot of sports marketing, after all, is driven by the desire to associate with a winner. In San Antonio, if the fans consistently left disappointed, it's unlikely that AT&T would slather its name and logo on most of the surface area of the arena or that Budweiser Select, Sprite, "your Texas Ford dealers,"

Southwest Airlines, and other sponsors would underwrite T-shirt giveaways, Bobble Head Night, and a halftime shooting contest.

By extension, the leagues have an incentive for the home teams to win. Although attendance and revenue rise in step with winning percentage for most teams, they rise even more sharply with *home* winning percentage. And healthier individual franchises make for a stronger collective. Does this mean leagues and executives are fixing games in favor of home teams? Of course not. But does it make sense that they would want to take subtle measures to endow the home team with (legal) edges? Sure. It would be irrational if they *didn't*.

The fact that the home field advantage exists is undeniable. But *why* does it exist? It's not for the reasons you might think. Let's start with conventional explanations and see where they fail us.

Conventional Wisdom #1: Teams win at home because of crowd support.

Let's go back to that February night in San Antonio. With a few minutes remaining before the tip-off, the public address announcer presented the starting lineups. The Blazers were introduced in a lifeless and staccato monotone that recalled the no-purchase-necessary-void-where-prohibited-consult-your-doctor-if-erections-last-more-than-four-hours-nobody-is-listening-to-me disclaimers at the end of commercials. Five Blazers came onto the court to a smattering of boos and then retreated into a team huddle.

Then it was time to introduce YOURRRRRRRRRRR SAN ANTONIOOOOOO SPURSSSSSSS!!! The lights dimmed. Strobes circled the floor. Music blasted. The indifferent PA announcer suddenly transformed himself into an unnaturally enthusiastic basso profundo as TOOOOOOoooooo-NNNNNNEEEEEeeeePARrrrrrK-ERrrrrrrrrr and his teammates were introduced. As the players took the floor to thunderous applause, voluptuous dancers with black-and-silver skirts aerosoled onto their impossibly sculpted bodies did elaborate pirouettes. Charles Lindbergh was barely treated to this kind of fanfare when his plane touched down in Paris.

On the first possession of the game, the Blazers' best player, Brandon Roy, missed a three-pointer from the baseline. As the Spurs headed up court, Parker commanded the ball. Slaloming around the Blazers' defense, he finally unspooled an elegant finger roll with his right hand and simultaneously was fouled by a late-arriving defender. The shot went in, triggering another robust "TONY PARKER" from the announcer. Parker made the free throw, and as the organ played, the crowd went nuts. So it went. Amid chants of "De-fense" the Blazers missed shot after wayward shot. Amid exhortations of "Da-da-da-da-da-daaaa . . . charge!" the Spurs made basket after basket. After barely two minutes, the Spurs were winning 7–2 and 18,672 fans were ecstatic.

It all stood to reason, right? It's logical that you will play better when you're being cheered and applauded and serenaded with chants, when your favorite songs blare on the PA system, and when your pregame introduction is accompanied by fireworks, figuratively and, as is sometimes the case in the NBA, literally. Conversely, common sense suggests that you will perform worse at your job when throngs of strangers are booing you and questioning the chastity of your sister and thwacking those infernal Thunderstix as you shoot free throws or strain to be heard in the huddle. And if chanting and taunting and noise don't bother you, athletes who have visited Philadelphia can attest that getting pelted with batteries while you play or cheered sarcastically when you're temporarily paralyzed doesn't exactly optimize performance, either.

But although the Spurs were outplaying the Blazers, it probably wasn't because of the crowd, the splashy introductions, or even the gyrating Silver Dancers—not directly, anyway. We've found that fans' influence on the players is pretty small. Much as crowds like to think they're vitally important in spurring on their team—the "sixth man," as they say in basketball—they're not. All those fans with their faces painted and their "number one" foam fingers pointing skyward? The Duke University student section with their clever taunts and Speedo attire? Despite Coach K's insistence to the contrary, they don't, sad to say, have much impact on the players.

How do we know this? One of the problems with testing the

effect of crowd support is that almost every feat is a function of not only the player and the crowd but also the defender, other teammates, the defender's teammates, and the referee. How do we isolate the crowd effect from all these other potential influences on the player? We need to look at an area of the game divorced from all these factors, such as free throws. Free throws are an isolated interaction between one player—the shooter—and the crowd that is trying to distract and heckle him. Also, all free throw shots are standardized; they are taken from the same distance of 15 feet at a basket standing 10 feet high regardless of where the game is played.

Over the last two decades in the NBA, including more than 23,000 games, the free throw percentage of visiting teams is 75.9 percent and that of home teams is . . . 75.9 percent—identical even to the right of the decimal point. Are these shooting percentages any different at different points in the game, say, during the fourth quarter or in overtime, when the score is tied? No. Even in close games, when home fans are trying their hardest to distract the opponents and exhort the home team, the percentages are identical. Sure enough, as sluggishly as the Blazers played in San Antonio, they would make 15 of their 17 free throw attempts (88.2 percent) even with fans behind the basket shouting and waving. The Spurs, by contrast, would make 75 percent of their attempts. Evidence of the crowd significantly affecting the performance of NBA players is hard to find.

What about other sports? In hockey, there's a rough equivalent to free throws we can use to gauge the crowd's potential influence on players. Beginning in the 2005–2006 season, the NHL adopted the "shootout" to settle ties in regular season games if the game remained tied after the standard overtime period. In a shootout, each team chooses three players to shoot one on one at the goalie. (Tournament soccer has a similar procedure with penalty kicks at the end of a tied game.) The team with the highest number of goals scored wins.

In the 624 games decided by shootouts in the NHL from 2005 to 2009, home teams won 308 (49.4 percent) and away teams won

316 (50.6 percent). In other words, for shootouts—held during clearly important times in the game when you'd expect the crowd to be *especially* involved and boisterous—the significant home ice advantage normally present in the NHL *evaporates*. When playing at home, shooters are no more successful than they are on the road. And when they're not successful, it is not because goalies are better at blocking shots at home as opposed to on the road, either. In a shootout, shooters are successful 33.3 percent of the time at home and 33.5 percent on the road, and goalies stop 51.5 percent of shots at home and 51.6 percent on the road. (About 15 percent of the time both home and away shooters miss the goal entirely.) If hockey fans aren't adversely affecting opposing players—or having a beneficial impact on the home team—during the most tense moments in a tie game, isn't it safe to assume that their support isn't affecting much when it's, say, midway through the second period? (There is a similar disappearance of home field advantage in tournament soccer penalty kicks.)*

In football, we could look at punters or kickers, who aren't exactly in total isolation from the rest of the players on the field, but pretty close. It turns out that yards per punt are identical for home and visiting punters (about 41.5 yards). Likewise, field goal success from the same distance and extra point accuracy are identical for kickers at home and on the road (about 72 percent on average).

Of course, punters and kickers are just two players, and you could question whether either has the ball long enough to be affected by a rabid crowd. Fair enough. Unfortunately, there isn't another isolated activity within the game of football we could point to in order to measure crowd influence outside everything else going on in the game. Instead, we could look at a number of offensive and defensive statistics to see where home teams fare better than visitors, recognizing that these advantages could

---

* Although shootouts occur only after the game has been tied and hence the two teams are evenly matched, implying that a 50–50 split of shootouts should be expected, the same could be said of overtime periods. In overtime, the teams also enter tied, yet the home ice advantage is still present in overtime.

come from a host of sources, including the crowd. It turns out that NFL teams rush better at home. They rush more—no surprise there since home teams are ahead more often—but also gain more yards per rushing attempt. Visitors, by contrast, pass more than the home team because they are usually behind and need to make up points in a hurry. But interestingly enough, visiting teams pass slightly *better* than home teams. (Who knew? NFL quarterbacks are a little better on the road than at home.) Though we might speculate that extreme crowd noise distracts visiting quarterbacks and makes their commands inaudible to their teammates, it doesn't seem to affect their performance. Thus, at least with respect to this aspect of the game, it's hard to say crowd noise contributes to the home team's success.

In baseball, the closest we can come to measuring the crowd's influence is to examine the pitcher. Not his ball-strike count—influenced, as it is, by the batter, the umpire, and the game situation—but his velocity, movement, and placement. We got the data thanks to the MLB.com technology Pitch f/x. A computer generates the location of the pitch, the height of the ball when released from the pitcher's hand, the speed at which the ball travels when it leaves the pitcher's hand as well as when it crosses the plate, and the degree to which the ball's direction changed or diverged from its path to the plate, both horizontally and vertically. Baseball researchers and Sabermetricians have been busily gathering and applying the data to answer all sorts of intriguing questions: Who has the nastiest sinker? How is Red Sox ace Josh Beckett's fastball different from his breaking ball? (For the record, his changeup has as much movement as his fastball, a big factor in his success.)

We obtained the last three years of these data, covering more than 2 million pitches, to answer a different question: Do pitchers actually pitch differently at home versus away? Before the Pitch f/x data existed, you couldn't really answer this question. You could only ask if the outcomes—balls or strikes—from pitchers were any different at home versus away? But again, outcomes are dependent on pitch selection, hitter reaction, umpire response, and game situation. Now, with the Pitch f/x data, we can simply ask: When Josh Beckett throws a fastball at Fenway, does it have more velocity, more movement, and better placement than it does when he is on the road?

Much like NBA players shooting free throws, on average Major League pitchers are equally accurate at home as on the road, throwing a ball within the strike zone 44.3 percent of the time at home and 44.5 percent of the time on the road. Pitches more than 1.5 inches outside the strike zone occur just as frequently at home as on the road (46.5 percent of the time). Even the most extreme pitches, those way out of the strike zone or those that hit the dirt, occur no more frequently on the road than at home.* In addition to having identical accuracy at home and on the road, pitchers throw with the same velocity—87 mph on average when the ball crosses the plate—and movement no matter where they play. We repeated these numbers for the same kind of pitch, categorized by Pitch f/x into changeup, fastball, curveball, four-seam fastball, split-fingered fastball, cut fastball, sinker, slider, and knuckleball, and found no difference in speed, movement, or placement for the same type of pitch by the same pitcher at home versus on the road. We tested the first inning versus later innings. Again, there was no difference. We tested different pitch counts. No difference. We even tested different game situations. Again, no difference. Pitchers appear to pitch no differently along any dimensions we can measure at home versus on the road, suggesting that neither the crowd nor the optics of the stadium influences their performance.

We can also use the Pitch f/x data to help gauge whether playing at home has any impact on batters. Controlling for all the factors that might make a batter hit better at home is nearly impossible, but we can control for a lot by looking at identical pitches—same speed, location, movement—in identical situations, home versus away. Does David Wright, the New York Mets' star

---

* If you were curious, pitches in the dirt occur 1.5 percent of the time at home and on the road.

third baseman, hit a waist-high, middle-of-the-plate, 90-mph fastball on a two-two count better at home, where the crowd is supporting him, than on the road, where he might be getting booed?

Location may be the key to real estate, but when it comes to hitting a baseball, for the average Major Leaguer geography doesn't matter. When a player swings at a pitch in the strike zone, his probability of hitting the ball is exactly the same at home and away. For pitches outside the strike zone, batters also fare equally at home and on the road. They are no more likely to swing at the same pitch at home than on the road and demonstrate no greater ability to swing at better pitches at home.

Of course, it could be the case that the crowd influences other aspects of the game—pumping up home teams to play better defense, encouraging them to greater effort. We can't rule this out, nor can we measure or isolate the crowd effect on these aspects of the game from other things going on at the same time. What we *can* observe is that in situations in which many of these other influences are "turned off" or controlled for, the crowd seems to have no effect. If the crowd is ineffectual in these isolated situations, it is at least questionable how much of an effect it could have in other situations.

Hey-batter-batter-batter-swing? Sorry, but he's going to do it equally well whether you're chattering or not. Just as he's going to shoot free throws, kick field goals, and deke the goalie comparably well whether you're encouraging him or cursing him.

Conventional Wisdom #2: Teams win at home because the rigors of travel doom the visitors.

By the end of the first quarter, the Spurs led comfortably, 29–20. The Blazers were shooting poorly, were committing scads of mental errors, and were conspicuously sluggish on defense. They were late to rotate, made only halfhearted efforts to block shots, and generally treated the lane as if it were a zebra-striped crosswalk: *No, really, after you. Go right ahead!* Portland's coach, Nate McMillan, would later observe that his players were "a step slow."

Who could blame them? The Blazers had played in Houston the previous night, boarded a plane, landed in San Antonio after midnight, and then taken a bus from a private airstrip to the hotel. Some players reported that they didn't fall asleep until dawn. As anyone who's taken a red-eye flight can attest, the grind of travel—the sitting, the dislocation, the time changes—can exact a steep price on the body. All the more so when the itinerary mirrors that of an NBA team pinballing randomly to Indianapolis one night, to Phoenix the next, and over to Dallas a few nights later. Your circadian rhythms are thrown off; your immune system can betray you. At the hotel, everything from the lumpy pillows to the inevitably ill-timed knock from the minibar stocker to the inadvertent wake-up call militates against a good night's sleep. It makes sense that the athletes who slept on their favorite pillows and woke up in their own beds that morning will outperform the ones who flew in earlier that morning. Particularly so for a team such as Portland, which—thanks to the Seattle Sonics relocating to Oklahoma City—is a significant flight away from every other team in the league.

We submit, however, that the travel doesn't much matter. The rigors of the road exist, but they don't underpin the home court advantage. Why do we say this? Consider what happens when teams from the same (or a nearby) city play each other, when the Los Angeles Lakers play the Los Angeles Clippers in the NBA—the two teams share the same arena—or the New York Rangers play against the New York Islanders or the New Jersey Devils in the NHL. For these games, the "rigors of travel" are nonexistent. Everyone is in his natural time zone and sleeping in his own bed. Yet if you look at all these "same city" games, you find that home teams have the exact same advantage they do in all the other games they host. Likewise, road teams don't lose more often when they travel greater distances. Controlling for the quality of the opponent, the San Antonio Spurs, for example, fare no better when they take puddle-jumpers to play the Dallas Mavericks and Houston Rockets than when they make longer trips to Boston, Toronto, and Miami.

We can take this one step further in the NHL, looking at games that involve not only long travel distances but also border crossing—which can require negotiating customs and other procedures that generally increase the pain of transit—by examining U.S. teams that play in Canada and vice versa. Yet we find no abnormal home ice advantage for U.S. teams visiting Canadian teams or vice versa, even for those farthest from home.

In Major League Baseball the rigors of travel aren't a significant issue, either. Just as in the NBA and NHL, for games involving teams from the same metro area—interleague play between the Chicago Cubs and White Sox, New York Yankees and Mets, Los Angeles Dodgers and Angels, San Francisco Giants and Oakland A's—the home teams win at *exactly* the same rate at which they normally do. We also know that home field advantage has been remarkably constant over the last century; it was virtually the same in MLB from 1903 to 1909 as it was from 2003 to 2009. This suggests that the teams jetting on chartered flights with catered meals, high-thread-count linens, and flat-screen televisions have no more success than did the teams that traveled to games in Pullmans and buses. (Either travel isn't causing the home advantage or teams need to rethink their jet purchases and the on-flight catering.)

Nor do the rigors of travel play much of a role in the NFL's home field advantage. Teams play only one game per week and in fact usually depart for a game a few days in advance to acclimate. As with the other sports, when nearby teams play—Oakland Raiders versus San Francisco 49ers, New York Giants versus New York Jets, Baltimore Ravens versus Washington Redskins—the home field advantage holds firm at its normal level.

Finally, we noticed that home field advantage in soccer is the same in countries such as the Netherlands, Costa Rica, and El Salvador, where travel distances are minuscule, as it is in countries as vast as the United States, Russia, Australia, and Brazil. It is yet another indication that travel isn't much of a factor.

Conventional Wisdom #3: Teams win at home because they benefit from a kinder, gentler schedule.

By halftime, the Blazers had whittled the San Antonio lead to three points. If both teams appeared fatigued—lacking "fresh legs," to use the basketball vernacular—it was with good reason. They had both played a game the previous night.

Because of the physical demands of running up and down a court for 48 minutes, it's exceedingly difficult to compete at full strength on consecutive nights. On the second night of back-to-back games, NBA teams win only 36 percent of the time. It was Charles Barkley who once referred to second games as "throwaways." In his inimitably candid way, he once explained, "You show up because they pay you to show up. But deep in your belly, you know you ain't gonna win."

Okay, we've discounted the effect of the "grueling" travel. But what about the fact that *visiting teams* play the vast majority of back-to-back games? Could that influence the home court advantage in the NBA? We think it does. And this particular Spurs-Blazers game notwithstanding, the vast majority of back-to-back games are played by road teams. Of the 20 or so back-to-back games NBA teams play each season, an average of 14 occur when they're on the road. That alone affects the home court advantage in the NBA. By our calculations, you are expected to win only 36 percent of those 14 games relative to your normal chances of winning on the road when you aren't playing back-to-back games. That translates into one or two additional games you will lose each season on the road because of this scheduling twist. In other words, home teams are essentially spotted an advantage of one or two games relative to road teams just from the NBA's scheduling of consecutive games.

It's not just the back-to-back games. Home teams not only play fewer consecutive games but also play fewer games in general within the same time span, such as the last three days or the last week or even the last two weeks. All this takes its toll on visitors. We estimate that about 21 percent of the home court advantage in the NBA can be attributed to the league's scheduling. Adjusting for this scheduling effect, the home court advantage drops down to 60 percent. So part of the explanation for the very high

NBA home court advantage is the way the league is arranging the schedule.

Recently, there were internal discussions—ultimately fruitless—in the NBA about reducing the season from 82 games to 75 games. The first games that would have been cut? The notorious back-to-back road games.

The league will tell you that the bunching together of a team's road games in as few days as possible is done to economize on travel. But by accident or design, it has the effect of working to the detriment of teams on the road. And it's not the only evidence suggesting that the league prefers to see home teams win. For instance, in home openers, the majority of teams start the season against weak opponents who had either inferior or similarly poor records the previous season. Just look at the Sacramento Kings and Washington Wizards, the two worst teams in 2008–2009; they started the 2009–2010 season playing five road games between them.

Says one NBA owner: "If only [fans] knew how the NBA scheduled games. Teams submit blocked dates for their arena [i.e., dates when the circus is using the building or the NHL team is using the facility]. The NBA picks 'marquee TV match-ups,' and then one guy figures out the rest with marginal help from software. Teams kiss his ass because we know he can throw more losses at us than Kobe can!"

To test the role of economic incentives, the most valuable NBA franchises, according to numbers from *Forbes*, are afforded a slightly stronger scheduling bias, as are teams in big markets. Yes, all teams play more consecutive games on the road than at home, but it's less so for the most valuable franchises in the biggest markets.

Remember how in most seasons *every* NBA team, even the Clippers, fares better at home than on the road? One look at an NBA schedule, and it starts to make sense why that is the case. When teams leave the comforts of home, they get hammered by the schedule makers, playing as many as three games in four nights, seldom in any logical order.

Seen through this lens, the action unfolding during the Portland–San Antonio game becomes clearer. The Blazers not only had played the night before at Houston (and lost) but had played four games in the week leading up to their game with the Spurs with only one day of rest between games. The Spurs meanwhile had played only three games the previous week and hadn't traveled in the last four days. If you take our numbers on back-to-back road games and factor in that Portland had also played one more game than San Antonio in the last week, the Blazers' chances of winning fall to less than one in three, which puts it in the vicinity of "ain't no way we're winning this motherf——" territory.

What about other sports? As in the NBA, teams in the NHL are brutalized by back-to-back games, which also occur disproportionately on the road. As physically demanding as the NBA is, the NHL may be even more taxing. In a typical season, road teams will play six more consecutive games than home teams do, which translates into about one or two extra home victories per team per season.

Scheduling is less of an issue in baseball; the 162-game schedule is set up so that teams play in three-, four-, and five-game series. When teams travel, they get to stay put in the visiting city, and the consecutive games have less of a physical effect on the athletes. The player who exerts himself the most—the pitcher—plays only once every five games.

In the NFL, the one league that publicly and unapologetically strives for parity—"any given Sunday" your team could beat the other team—there is virtually no evidence of scheduling bias. Even in home openers, the most successful NFL teams are not favored. In fact, we find the opposite: NFL teams that did well the previous season are more likely to face a *better* opponent in their opening home game than they are to face a team that did poorly last season.

■ ■ ■

By contrast, in college sports, scheduling plays a *huge* role in the home team advantage. College boosters would have you believe that the exceptionally high winning percentage in NCAA sports is a consequence of rabid school spirit, the pep bands and cheerleaders, and those exuberant undergrads annoying opponents with witty cheers and taunts. But most followers of college sports are likely to guess what's really driving a large part of the high home field advantage. It's the scheduling of weak opponents—cupcakes, patsies, sacrificial lambs, road kill, call them what you will—early in the season.

Although the NCAA and the conferences set the schedule for most "in-season" games, the individual schools are generally free to negotiate their own preseason schedules. At large schools, there is an incentive to pad teams' records early in the season. Stacking the scheduling deck in their favor, teams from the six "big" football conferences—the Big Ten, Pac-10, SEC, ACC, Big 12, and Big East—win almost 90 percent of their home openers. In addition to pleasing the crowd, especially those cotton-head donors sitting in the prime seats, early success bolsters the team's chances of reaching the postseason bowls and tournaments, which come with a direct financial payoff and, generally, a spike in alumni contributions.

At small schools, there are incentives to play along. One is to raise revenue. Often, playing at Big State U ensures a monetary reward far superior to what the team could have earned playing a smaller opponent. In 2006, for instance, the small-time football program at Florida Atlantic University was paid $500,000 to play at Clemson in the season opener. FAU then reportedly made an additional $1.325 million playing its next three games *at* Kansas State, Oklahoma State, and South Carolina. From those four road games alone, they covered a sizable chunk of the annual operating expenses for their entire athletic department. But they lost the four road games by a combined score of 193–20.

Less cynically, like any underdog, small schools also thrill at the chance to "make a name for themselves" and elevate their profile on the off chance that the team can spring an upset. (Who had even heard of the small Hawaiian school Chaminade before its basketball team's momentous upset of a Ralph Sampson–led top-ranked Virginia team in 1982?) Even in defeat, the small school usually appears on national television, a big draw to potential recruits. Plus, the players leave with a sense of how far they are—sometimes it's not very—from the next level of competition. In short, everybody usually benefits from this arrangement.

It's worth noting that scheduling a "patsy" opponent can back-fire financially, sometimes spectacularly. In 2007, the University of Michigan football team (ranked fifth in the nation at the time) lost its home opener to tiny Division I-AA Appalachian State. Not only did such a disgraceful defeat ruin any chance of Michigan playing in a big (read: well-paying) bowl game regardless of what happened the rest of the season, but such an embarrassing loss—at home!—surely had an effect on alumni donations.

We found that home schedule padding accounts for roughly half of the home team advantage in college football. If we adjust for the quality of teams—or look at in-conference games, where the conference and not the big schools sets the schedule—home team winning percentage drops from 64 percent to 57 percent. Incredibly enough, that 57 percent is almost the exact same rate at which the home teams win in both the NFL and Arena football. For college hoops the numbers are similar. Of the impressive 69 percent home court advantage in NCAA basketball, a little more than half can be explained by early-season schedule padding. Accounting for these scheduling biases and strength of opponent, the home advantage in college basketball declines to 63 percent, the same as in the NBA.

But scheduling bias gets us only so far. It accounts for half of the home field advantage in college sports; it partially explains the home field advantage in the NBA and NHL. In Major League Baseball and the NFL—and, as it turns out, in soccer as well—it doesn't explain it at all.

Conventional Wisdom #4: Teams win at home because they are built to take advantage of unique "home" characteristics.

Fulfilling the Portland coach's prophecy, the Spurs pulled away in the second half and beat the Blazers handily, 99–84, in a contest that was virtually uncontested. From the start, the Blazers competed with no urgency or passion, missing nearly two-thirds of their shots and giving only periodic consideration to defense, as if resigned to defeat. The postgame locker room hardly called to mind the picture of despair. Just one more game in Minnesota and the road trip would be over. "A few more days, man," said the Blazers' center, Greg Oden, who hadn't even played that night. "A few more days and we get to go home."

As for the Spurs, they played generally unimpeachable basketball. They competed capably, defended capably, and shot the ball well. Unmistakably, the star of the game was Parker, who darted around the court, scoring 39 points in 35 minutes. In a battle of images, Spurs coach Gregg Popovich gushed that Parker was "a superstud again." Nate McMillan, the Portland coach, likened Parker to "a roadrunner blowing by us."

Sterlingly as Parker had played, he was, by his own admission, at his best when paired with Tim Duncan. In addition to coaching the team, Popovich was the team's chief architect, and his decision to draft Parker as a complement to Duncan's inside presence was a coup that had paid immense dividends—not least multiple NBA titles. The twenty-eighth selection in 2001, Parker represents one of the great steals in recent NBA drafts. When making personnel decisions, Popovich told us that he considers dozens of factors: How will the players fit into the tapestry of the team? How much will they cost? How will they feel knowing they'll be operating in the considerable shadows of the Spurs' three stars? How will they acclimate to a "small market" that lacks the beaches of Los Angeles and the nightlife of New York and Miami?

What he doesn't worry about is how they will play specifically in the AT&T Center. From arena to arena, the baskets are 10 feet off the ground and 94 feet apart, 15 feet from the free throw line. Throughout the NBA, the playing surface is standardized. The games are always played indoors in climate-controlled venues. Even the placement of the decals on the court must conform to league regulations. It's the same in the NHL. For all intents, a rink is a rink is a rink.

In the NFL, each field is 120 yards long, including the end zones, and roughly 53 yards wide; but the climate and playing conditions can vary immensely. A December game in San Diego, California, is played in a much different environment than a December game in Buffalo, New York. Is the home field advantage in football influenced by teams tailoring their rosters to the weather?

We didn't find that to be the case.

Much as broadcasters talk about those poor teams from the tropical precincts—say, the Miami Dolphins and San Diego Chargers—faltering in the "frozen tundra" of Lambeau Field in Green Bay or thermally challenged lakeside stadiums in Buffalo and Cleveland, climate, we've found, is largely irrelevant. If NFL teams are built to take advantage of their home weather, we should observe cold weather teams winning disproportionately in cold weather games. We also should observe teams playing poorly when venturing to markedly different climates. Finally, we should observe that "domed teams," because they play in climate-controlled chambers, are simply insensitive to home weather. Think of this last situation as a placebo test. If the control group experiences similar reactions even though they weren't administered any treatment or medicine, you know it wasn't the drug but something else that caused their reactions. Similarly, if domed teams seem to perform differently depending on the outside weather, it must be something else influencing the results. For example, just as the weather tends to get worse late in the NFL season in December, teams may play better at home late in the season, having nothing to do with weather at all.

After studying data from every NFL game from every season between 1985 and 2009—nearly 6,000 games—and matching those games to the outside temperature and wind, rain, and snow conditions, we found that cold weather teams* are no more likely to win at home when the weather is brutally cold, nor are warm weather teams more likely to win at home when the temperature

---

* Cold weather teams are Buffalo, Pittsburgh, New York Giants, New York Jets, Philadelphia, Washington, Baltimore, Green Bay, Chicago, Denver, New England, Cleveland, Cincinnati, and Kansas City. Dome teams are obvious, and the rest are considered warm weather teams. Note, too, that there are several teams from a cold weather climate that play in domes: Minnesota, Indianapolis, St. Louis, and Detroit. We also adjust for the prior winning percentage of each team in each game to control for team quality.

is awfully hot. And the home winning percentages for dome teams immune from extreme weather conditions—our placebo test—do not vary with the weather any more than they do for cold and tropical weather teams. Even looking at the most extreme cases, when a warm weather team has to play in extremely cold weather or a cold weather team plays in humid and hot conditions, there is little to no unusual effect. Contrary to conventional wisdom, weather gives a team no additional home advantage. Either teams are not built to suit their home weather conditions or, if they are, it doesn't seem to have much effect on the outcome of games.

What about baseball? After all, not only do the playing conditions vary but—one of the sport's great appeals—each stadium is unique. Don't the home players have an advantage, as they're more familiar with their ballpark's idiosyncrasies? Don't teams stock their rosters with players who are better suited for their park's features? And couldn't this influence the home field advantage in baseball?

Yes and no.

There's no question that the Boston Red Sox, for instance, have an advantage playing at Fenway Park. The Sox outfielders know the Green Monster, the notorious 37-foot-high left field wall, the way Thoreau knew Walden Pond. Unlike the opponent, they're well acquainted with caroms and angles and the effect of the wind. (There's even an ersatz Green Monster scheduled for construction at the Red Sox spring training facility so the organization's minor leaguers can familiarize themselves with the wall's distinctive features.) Similarly, Sox hitters know that although the Green Monster is high, it's also deceptively shallow—barely 300 feet from home plate—and they adjust their swings accordingly. Surely this has an effect on Boston's home winning percentage.

You might also surmise that baseball players are familiar with the unique optics of their home park. The home batters see the ball better; the visiting pitchers exhibit less control. But we already know that once we've controlled for other factors—pitch count, game situation, and so forth—players don't hit the ball appreciably better at home and pitchers don't throw appreciably less accurately on the road. Thus, that can't be the reason home teams win more games.

What about the notion that baseball teams win more games at home because they tailor their rosters to the idiosyncrasies of their ballparks? The teams that play in parks with, say, shallow right field porches recruit more left-handed hitters. The teams with uncharitable dimensions recruit superior pitchers and speedy outfielders. How much does this affect the home field advantage?

Since it would be impossible to consider every ballpark and how different types of players might be better suited to each, we looked at the most obvious case in baseball and the one likely to have the biggest impact: "hitter-friendly" ballparks versus "pitcher-friendly" ballparks. The Sabermetrics community helped us identify which ballparks historically were hitter-friendly, using total number of runs, hits, extra-base hits, and home runs produced in each ballpark by all teams each season. We then asked: Do teams from hitters' parks outhit their visitors by more than teams from pitchers' parks do? If teams from hitters' parks are being stacked with sluggers, we should see them outhit their visiting opponents by a wider margin than that of teams from pitchers' parks. Yet we don't. Teams from hitters' ballparks outhit their visitors by the same amount as home teams in pitchers' parks do—same differences in batting average, home runs, doubles, triples, slugging percentage, and runs created. We even found this to be the case when a hitting team plays host to a team from a pitchers' ballpark, where you'd expect the widest difference.

We also looked at how teams from hitters' ballparks play *away* from home. If their lineup is stocked with hitters, they should hit better than other teams no matter where they play. (Plus, by looking at other ballparks we also remove any other home advantages, such as crowd, familiarity with field of play, and travel.) But we found that teams that play at home in hitters' ballparks hit no better on the road than teams that play host in pitchers' ballparks do when they're on the road, even going so far as to control for the same stadium. That is, the Colorado Rockies (who play in a hitters' park) hit as well as the New York Mets (who play in a pitchers' park) when they each play in Busch Stadium in St. Louis.

All this evidence indicates that either teams aren't stacking their rosters to suit their home stadiums or, if they are, it's not making much of a difference. Bear in mind that the home field advantage in sports is lowest in baseball. Even if "roster tailoring" is a factor in some cases, it doesn't get us very far in explaining the home field advantage phenomenon overall.

We should add that deception and "dark arts" don't seem to be much of an explanation for the home advantage, either. In past eras, it was different. In 1900, for instance, a shortstop on the visiting Cincinnati Reds noticed that the Phillies' third-base coach stood in a puddle each inning. When the shortstop investigated, he found that under the puddle was a wooden box. It turned out that a Phillies backup catcher sat in the outfield bleachers armed with high-powered lenses and stole signs from the visiting team. Then, using a buzzer that was connected to the wooden box with wires that ran under the field, he used Morse code to convey the pitch to the third-base coach. The coach then relayed the information to the batter. Little wonder the Phillies won two-thirds of their games at home and fewer than half on the road.

Through the years, other home teams have used elaborate plots to steal signs from the visitors. For years, home groundskeepers in baseball would water the field into a bog when speedy visiting teams were in town. The Boston Celtics were notorious for jacking up the heat in the visitors' locker rooms so that halftime resembled a session in a sauna. The University of Iowa football team once ordered the visiting locker rooms painted pink, hoping it would make the opponent feel passive or emasculated. It's unclear if any of this worked—and it probably didn't—but because of the standardized league rules, the stiff deterring punishment for cheating, and surveillance technology, it would be hard to pull off this kind of skulduggery today.

■ ■ ■

So let's take stock of all we know: When athletes are at home, they don't seem to hit or pitch better in baseball, shoot free throws better in basketball, slap goals better in hockey shootouts, or pass better in football. The home crowd doesn't appear to be helping the home team or harming the visitors. We checked "the vicissitudes of travel" off the list. And although scheduling bias against the road team explains some of the home field advantage, particularly in college sports, it's irrelevant in many sports. The notion that teams are assembled to take advantage of unique home characteristics isn't borne out, either.

Yet if home teams are winning more games so consistently, players on those teams surely must be doing *something* better than their opponents. What else is giving the home team its sizable edge?

■ ■ ■

Thanks to the quirks of the NBA schedule, the Blazers and the Spurs played again four nights later. Duncan, San Antonio's exceptional big man, had regained his health and was back in the lineup. If the Spurs had beaten the Blazers by 15 points when he was on the bench, surely they would crush them with Duncan in the game. Right?

But this time the two teams played in Portland. This time the Blazers would have the exuberant PA announcer, the dance teams, the 20,000 partisan fans. This time the Blazers shot more free throws. This time the Spurs committed more fouls and turnovers. This time the Blazers won 102–84, a whopping 33-point swing from their game only 96 hours earlier.

Maybe these athletes and coaches are right, after all, to adopt a defeatist attitude when heading off on the road.

But why?

SO, WHAT *IS* DRIVING THE HOME FIELD ADVANTAGE?

Hint: Vocal fans matter,
but not in the way you might think

It had the makings of a nearly perfect day. Jack Moore had just finished his sophomore year at the University of Wisconsin and was home for a few summer weeks, living with his folks. Marooned in the Mississippi River town of Trempealeau, Wisconsin, Jack was blissfully free of pressure, with generous rations of free time. He had a job coaching baseball, but the games didn't start until the evening. On this Friday of the 2009 Fourth of July weekend, Jack's beloved Milwaukee Brewers were playing an afternoon road game against their rivals the Chicago Cubs.

Air-conditioning blasting, Jack flicked on the cable to the regional sports network and sat down on the couch to watch. The Brewers were coming off a magical 2008 season in which they won 90 games and reached the playoffs. In the off-season, Milwaukee's ace, C. C. Sabathia, was poached by the Yankees. It was the numbingly familiar fate of a small-market team: The Brewers had been unable to match New York's $161 million contract offer. Jack was okay with that. A math major, he knew the economic realities and understood why Milwaukee could not afford to retain a star at those prices. Besides, the Brewers' 2009 incarnation was easy to root for, a fun team with a winning record, filled with young and energetic players.

The game was a rare Wrigley Field pitching duel pitting the Cubs' ace, Carlos Zambrano, then a Cy Young Award candidate, against Milwaukee's veteran Jeff Suppan. The game was tied 1–1 after nine innings, which was all good with Jack, a former high school baseball player who was thoroughly capable of appreciating a low-scoring affair. "It was one of those games," he recalls, "that remind you why you like baseball so much."

Then, in the bottom of the tenth inning, Jack's idyllic afternoon was ruined. The Brewers had summoned Mark DiFelice, a right-handed pitcher who had recently won his first Major League game at age 32. When the Cubs loaded the bases, DiFelice faced Chicago's third baseman, Jake Fox, a utility man who'd ricocheted between the majors and the minors. With a full count, two outs, and the decibel level soaring at Wrigley Field, DiFelice threw four consecutive pitches that Fox fouled off. On the next pitch of the at-bat, DiFelice reared back and fired a cutter that froze Fox and shot past him. After an awkward pause, home plate umpire Bill Welke popped up from his crouch and . . . stood idly. Ball four. The winning run had been walked home: Cubs 2, Brewers 1.

The crowd goes wild. Jack Moore of Trempealeau, Wisconsin, goes ballistic. "For five minutes, I just screamed words you can't print," he says. "Anyone who knows baseball knew that was a strike." For years, fans in Jack's position would bitch and moan and dispute balls and strikes until last call. But this was 2009, and Jack wasn't interested in an argument; he was interested in a straight, objective answer. He fired up his Internet browser, logged on to MLB.com, and clicked on Pitch f/x. Sure enough, DiFelice's pitch was gut-high and clearly within the upper-inside part of the strike zone. Minutes after the game had ended, right there in his parents' home in small-town Wisconsin, a 19-year-old was able to confirm his suspicions. The ump had blown the call, permitting the home team to win.

■ ■ ■

What sports fan doesn't harbor a belief that the officials are making bad calls against his or her team? It's a home crowd that voices this displeasure the loudest. The criticism ranges from passably clever ("Ref, if you had one more eye, you'd be a Cyclops!") to the crass ("Ref, you might as well get on your knees because you're blowing this game!") to the troglodytic ("You suck!"). Dissatisfaction is voiced individually and also collectively, often in a stereo chant of "Bullshit! Bullshit!" In Europe—quaint, civilized Europe—there are even various soccer websites that enable fans to download antireferee chants as ringtones.

What we've found is that officials *are* biased, confirming years of fans' conspiracy theories. But they're biased not against the louts screaming unprintable epithets at them. They're biased *for* them, and the bigger the crowd, the worse the bias. *In fact, "officials' bias" is the most significant contributor to home field advantage.* "Home cooking," as it's called, is very much on the menu at sporting events.

A statement like that had better have some backing, and we're prepared to provide it. Warning: An assault of numbers awaits. But stick with us and we'll walk you through it. We think the payoff is worth it.

■ ■ ■

Let's start by determining how to measure ref bias. You could examine the accuracy of calls made by the officials and whether that accuracy differs for calls favoring the home team versus the away team. But doing that is a challenge because it requires a great deal of subjectivity as well as a deep knowledge of the circumstances of the game. Was it really a foul? Was it really pass interference? What else was happening during the game at that time? In light of the speed of the game and the reactions of players within the game, it is nearly impossible to control for all the potential factors that could lead to differing calls for the home and away teams.

Suppose we find that more fouls are called against road teams than against home teams—which, by the way, is often the case. Does this indicate a referee bias in favor of the home team? Maybe, but not necessarily. What if teams play more aggressively on the road? After all, road teams know that statistically, they are already more likely to lose. Or what if the road team, exhausted from those back-to-back games, lacks the energy for proper defense and clutches and grabs instead? They might be inclined to commit more fouls regardless of any referee bias, and so it's difficult to identify the *causal* factor. Are referees *causing* more road team fouls because of bias against the road team? Or are players causing referees to call more fouls because of more sloppy or aggressive play? Or is there a third factor causing both?

We looked for a component of the game the refs control that isn't influenced or affected by players. We found it in a sport for which we have not had much success in explaining its sizable home advantage—soccer. It also turns out that had it not been for a diligent grandmother from Spain religiously watching and recording years' worth of Sunday evening matches, we might not have discovered this bias at all.

In soccer, the referee has discretion over the addition of extra time, referred to as "injury time," at the end of the game to make up for lost time resulting from unusual stoppages of play for injuries, penalties, substitutions, and the like. This extra time is rationed at the discretion of the head referee and is not recorded or monitored anywhere else in the stadium.

As best he can, the referee is supposed to determine the accumulated time from unusual stoppages—itself a subjective measure—and add that time at the end of regulation. So does the referee's discretion favor the home team? If so, he would lengthen this time when the home team is behind at the end of the game and reduce it when the home team is ahead, extending or shortening the game to increase the home team's chances of winning.

Using handwritten notes that his elderly mother had gathered logging matches she'd watched from her living room in Spain, Natxo Palacios-Huerta, a London School of Economics professor, joined with two colleagues from the University of Chicago, Luis

Garicano and Canice Prendergast—all soccer fanatics—to study the officials' conduct during games. The researchers were, quite justifiably, struck by what they found. Examining 750 matches from Spain's premier league, La Liga, they determined that in close matches with the home team ahead, the referees ritually shortened the game by reducing the extra time significantly. In close games in which the home team was behind, the referees lengthened the game with extra injury time. If the home team was ahead by a goal at the end of regulation, the average injury time given was barely two minutes, but if the home team was behind by a goal, the average injury time awarded was four minutes—twice as much time. Sure enough, when the score was tied and it wasn't clear whether to increase or decrease the time for the home team, the average injury time was right around three minutes.

What happened when the home team was *significantly* ahead or behind? In games that were not close, there was no bias at all. The extra time added was roughly the same whether the home team was ahead by two goals or more or behind by two goals or more. This makes sense. A referee has to balance the benefit of any favoritism he might apply with the costs of favoritism—harm to his reputation, media scrutiny, and potential reprimands. Adding additional injury time when the score was so lopsided was unlikely to change the outcome and therefore accrue much benefit, so why do it and risk the potential cost?

The study also looked at what happened when, in 1998, the league altered its point structure from awarding teams two points in the standings for a win (and one for a draw and zero for a loss) to three points for a win. That change meant that a win was suddenly worth a lot more than it had been before and the difference between winning and tying doubled. What did this do to the referee injury time bias? It increased it significantly. In particular, preserving a win against the possibility of a tie now meant a lot more to the home team, and so the referees adjusted the extra time accordingly to reflect those greater benefits.

This wasn't unique to Spain. Researchers began looking for the same referee biases in other leagues—not hard given the global popularity of soccer. They found that the exact same injury time bias in favor of the home team exists in the English Premier League, the Italian Serie A league, the German Bundesliga, the Scottish league, and even MLS in the United States.

If referees are willing to alter the injury time in favor of the home team, what else might they be doing to help ensure that the home crowd leaves happy? We found that referees also award more penalties in favor of the home team. Disputed penalty shots and goals tend disproportionally to go the home team's way as well. Looking at more than 15,000 European soccer matches in the English Premier League, Spanish La Liga, and Italian Serie A, we found that home teams receive many fewer red and yellow cards even after controlling for the number of penalties or fouls on both teams. The dispensing of red and yellow cards has a large impact on a game's outcome. A red card, which sends the offend-ing player off the field, reduces a team's chances of winning by more than 7 percent. A yellow card, which precedes a red card as a stern warning for a foul and may therefore cause its recipient to play more cautiously, reduces the chances of winning by more than 2 percent. These are large effects. When a single yellow card, followed by a red card, is given to a visiting player, it means the home team's chance of winning, absent any other effects, jumps to 59 percent. Add the injury time, fouls, free kicks . . . and it sud-denly isn't so surprising that the home team in soccer wins nearly 63 percent of its games.

But could this be limited to the idiosyncratic world of European soccer? Surely, American sports wouldn't be subject to the same referee bias . . . would they?

Remember how, despite a significant home team advantage, athletes do not hit or pitch, shoot free throws, slap goals, or pass the football appreciably better at home than they do on the road? This prompts the question: What *do* home teams do better that allows them to achieve a higher winning percentage?

In baseball, it turns out that the most significant difference be-tween home and away teams is that home teams strike out less and walk more—a lot more—per plate appearance than do away teams. This could be for lots of reasons. One interpretation: Home team batters see the ball better or away team pitchers exhibit less control. But this contradicts our earlier results for batters and pitchers—in controlled, isolated environments, they hit and pitch the same at home as they do on the road. And as we've seen, road players in MLB aren't performing worse because they're exhausted from the travel.

Balls and strikes are the domain of the head umpire. Could the umpire be biased toward the home team? This would explain the differences in strikeouts and walks despite the lack of any difference in hitting and pitching.

But strikeouts and walks are not the right statistic to measure, because many strikes occur when a batter swings and misses or fouls off a ball. In such cases, there is no umpire discretion. A better metric to look at is *called* balls and strikes.* In other words, look only at pitches that do not involve swinging by the batter. It turns out that home batters receive far fewer called strikes per called pitch than away batters do.

It's even more apparent when we look at called strikes and balls at different points in the game. Certain situations have a much bigger impact on the game's outcome than others. Fortunately for us, Sabermetrics, an analysis of baseball through objective evidence, provides another useful tool to gauge the importance of a particular situation. A stats wizard, Tom Tango, devised a metric called the Leverage Index to measure the relative importance of any game situation. The idea is to take every game situation and consider every possible scenario that could occur in that situation, the likelihood of each scenario playing out from that point, and what effect each of those scenarios would have on the ultimate outcome of the game. Add up all these possibilities, their likelihood of occurring, and their potential impact on the game and you have a measure of how crucial the current situation is. A Leverage Index of 1 is the average situation; an index of 2 means

---

* Eliminating intentional walks.

the situation is twice as crucial. Here are two extreme examples: Down by four runs with two outs and nobody on base in the bottom of the ninth, where the game isn't in much doubt, translates into a Leverage Index of 0.1—the situation is one-tenth as crucial as the average situation. Down by one run in the bottom of the ninth with two outs and the bases loaded, where the game is on the line, gives a Leverage Index of 10.9. It is almost 11 times more crucial than the average situation.

Using the Leverage Index to examine called strike and ball counts in different situations, we found, just as with the soccer referees, that in low-leverage situations, when the game is not in much doubt, the home team advantage in receiving fewer called strikes and more balls goes away. But as the following chart shows, the called-strike advantage for home teams grows considerably as the game situation gets more and more important. In noncrucial and average situations, the home team receives about the same strike calls as, or even a few more strike calls than, the away team per called pitch, but that changes dramatically when the game is on the line. In crucial situations, the home team receives far fewer called strikes per called pitch than does the away team.

DIFFERENCE IN PERCENTAGE OF CALLED PITCHES THAT ARE CALLED STRIKES ON HOME VS. AWAY BATTERS

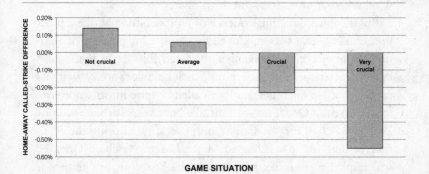

This makes sense. If the umpire is going to show favoritism to the home team, he or she will do it when it is most valuable—when the outcome of the game is affected the most. You might even contend that in noncrucial situations the umpire may be biased against the home team to maintain an overall appearance of fairness.

Think back to that Jake Fox pitch in the Cubs-Brewers game, on a 3–2 count with the bases loaded and a tie game on the line in the bottom of the tenth inning. It was an astronomically high-leverage situation. Knowing the statistics, you would have bet the house that the pitch wouldn't have been called a strike. And it wasn't.

Let's look at other calls that fall under the domain of the umpires, in particular, close calls that typically elicit a home crowd reaction. Two good examples would be stolen bases and double plays. We found that home teams are more likely to be successful when stealing a base and when turning a double play, yet the distance between the bases is identical in every stadium—stolen base success can't be driven by home field idiosyncrasies. In addition, the success rates of home teams in scoring from second base on a single or scoring from third base on an out—typically close plays at the plate—are much higher than they are for their visitors in high-leverage/crucial situations. Yet they are no different or even slightly less successful in noncrucial situations. (Third-base coaches, take note: If it's a close game and you're playing at home, windmill your arms and send the runner!)

But the most damning evidence of umpire bias might be a function of a tool that was employed for the specific purpose of eliminating umpire bias. Remember the Pitch f/x system that tracks the characteristics of each pitch, including location? Well, its predecessor—a digital technology called Umpire Information System (UIS) from QuesTec—was installed five years earlier by Major League Baseball for the specific purpose of monitoring the accuracy of umpires. According to Major League Baseball, QuesTec was implemented in six ballparks in the first year; by the time it was discontinued in 2008, 11 ballparks had the technology.* With two cameras positioned at field level and two in the upper deck, QuesTec combined the four images to track where the ball crosses the plate, and it was used by baseball executives to determine how closely an umpire's perception of the strike zone mirrored reality.

We also used the presence of QuesTec to evaluate umpire accuracy, but in a different way. We asked whether the same umpire behaved differently when he *knew* the cameras were monitoring him. If the home field advantage in called strikes disappears when the umpires know they're being watched—while everything else stays constant—it's pretty clear that official bias underlies it. Imagine you own a coffee shop and put out a jar in which patrons can donate or take loose change. You notice at the end of each day that the jar is empty. You deduce that either some customers are taking advantage by depleting the jar or your employees are stealing the coins. You tell only your employees that you are installing a hidden video camera. If the change jar is full at the end of each day, you're pretty darn sure it was your employees, not customers, who were to blame.

To test our theory, we first compared all pitches, about 5.5 million of them, from 2002 to 2008 made in stadiums using QuesTec versus those without it. For example, we looked at all called pitches when the Astros visited the Cardinals (at their non-QuesTec stadium) and when the Cardinals visited the Astros (at their QuesTec-equipped stadium).

What did we find? Called strikes and balls went the home team's way, *but only* in stadiums without QuesTec, that is, ballparks where umpires were not being monitored. This is consistent with an umpire bias toward the home team causing the strike-ball discrepancy. We also found something surprising. Not only did umpires

---

* According to Major League Baseball, the 11 franchises whose ballparks were equipped at various times with QuesTec were the Arizona Diamondbacks, Boston Red Sox, Chicago White Sox, Cleveland Indians, Houston Astros, Los Angeles Angels of Anaheim, Milwaukee Brewers, New York Mets, New York Yankees, Oakland A's, and Tampa Bay Rays.

not favor the home team on strike and ball calls when QuesTec was watching them, they actually gave *more* strikes and *fewer* balls to the home team. In short, when umpires *knew* they were being monitored, home field advantage on balls and strikes didn't simply vanish; the advantage swung all the way over to the visiting team.

We then looked at the same pitch counts in low-leverage (not crucial) and high-leverage (crucial) points in the game. Again, when a plate appearance is expected to have little effect on the outcome of the game, there is no bias for or against the home team. Umpires call things evenly whether QuesTec is present or not. But when the at-bat can have an impact on the game, we found both biases to be even more extreme. That is, when the game is on the line, home teams in non-QuesTec stadiums get a big strike-ball call advantage and those in QuesTec stadiums get a huge strike-ball call *disadvantage*.

In practical terms, when the umpire is *not* being monitored by QuesTec, a home batter in crucial game situations will get a called strike only 32 percent of the time if he doesn't swing. In the same situation, a batter from a visiting team gets a called strike 39 percent of the time. That's a big difference. Now consider the same two situations when the umpire *is* being monitored by QuesTec. Here the home batter gets a called strike 43 percent of the time, and the away batter only 35 percent of the time.

If we were consultants to a team equipped with umpire-monitoring technology, our first piece of advice would be: Get rid of QuesTec; it's wrecking your home field advantage. (How many teams would have agreed so readily to QuesTec if they knew these numbers?) Of course, if we were consulting for MLB, we might have encouraged them to install the technology in *all* the ballparks or at least tell the umpires that was the case. (Today, that's essentially what MLB has done.)

We also found the same results for the QuesTec stadiums before and after the system was installed. The called strike-ball differences between home and away teams declined sharply after QuesTec installation, and the decline was particularly pronounced in crucial situations. Even the *same* umpire behaved differently depending on whether QuesTec was present, calling more strikes and fewer balls on home batters when he was being monitored and doing the opposite when he wasn't.

Why would the home team advantage for strike-ball calls, particularly in crucial situations, switch completely in the other direction when QuesTec is present? You'd think the advantage would just disappear, creating no bias, but in fact the bias goes in the opposite direction. We suspect that as with the referees in soccer, umpires have to balance the costs and benefits of any bias (conscious or not) they might exhibit. If you know you are being monitored, you want to eliminate any perception of bias. And when the game is on the line, you know that any perceived bias will be scrutinized even more closely. With the speed of the game and the uncertainty of whether a 95-mph fastball hit or missed the outside corner of the plate, umpires may become overly cautious. Worried about accusations of home-team favoritism, the umpire seems to err in the other direction, particularly in situations that will be monitored and analyzed heavily afterward.

What about the other potential umpire biases we found that might benefit the home team, such as stolen bases, double plays, and other close plays? They remain the same whether QuesTec is present or not, which makes sense. After all, QuesTec monitors only the strike zone. It affects no other part of the game. Calls remain in favor of the home team because there's no "surveillance video" on those calls.

If QuesTec is our smoking gun in the case to prove umpire home team bias in Major League Baseball, Pitch f/x provides the ballistic support.* Using the Pitch f/x location data of the millions of pitches we examined earlier, we asked a series of questions: How likely is it that when the pitch is actually out of the strike zone, an umpire will call a strike on the home team versus the away team? How often is a ball called on a pitch actually within the

---

* Pitch f/x is now in every ballpark, and thus, one could argue, umpires are now monitored everywhere. However, Pitch f/x—unlike QuesTec—is not being used to evaluate performance. There's a big difference between casually monitoring umpires and bosses formally monitoring umpires.

strike zone when the home team is batting versus the away team, the situation that enraged the die-hard Brewers fan Jack Moore? What about pitches just in or just out of the strike zone? How do these calls change in critical situations?

The following chart graphs the difference in the percentage of called pitches that are called strikes on home versus away batters for pitches within three inches of the dead center of the strike zone, pitches way out of the strike zone (at least three inches), and pitches within 1.5 inches of the strike zone, for example, just on or off the corners. We report the numbers separately for two-strike, three-ball, and full counts.

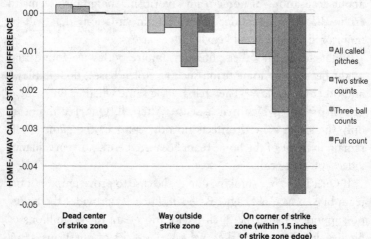

DIFFERENCE IN PERCENTAGE OF CALLED PITCHES THAT ARE CALLED STRIKES ON HOME VS. AWAY BATTERS

Note two points: (1) The home-away differences are largest for two-strike and three-ball counts and especially for full counts. (2) For the most ambiguous pitches—the ones on the corners—the home-away called-strike discrepancy is largest, which makes sense. The umpire has less discretion over pitches that are less ambiguous. Umpires will be reluctant to make a biased call if the pitch is obviously a strike. In fact, for pitches in the dead center of the strike zone, there is no bias at all. Umpires call these pitches correctly 99 percent of the time whether a home or a visiting batter is standing in front of them. For pitches way outside the strike zone, the umpire has a little more leeway and shows a slight bias in favor of home batters. The umpire has the most discretion for pitches on the corners, and there the home batter bias is largest.

Over the course of a season, all of this adds up to 516 more strikeouts called on away teams and 195 more walks awarded to home teams than there otherwise should be, thanks to the home plate umpire's bias. And this includes only terminal pitches—where the next called pitch will result in either a strikeout or a walk. Errant calls given earlier in the pitch count could confer an even greater advantage for the home team.

How much do these differences contribute to the home field advantage in baseball? Well, we need to know the value of receiving an extra pitch instead of striking out and the value of being awarded first base instead of facing another pitch, but here's a rough estimate. Taking the value of a walk and a strikeout in various game situations, this adds up to an extra 7.3 runs per season given to each home team by the plate umpire alone. That might not sound significant but cumulatively, home teams outscore their visitors by only 10.5 runs in a season. Thus, more than two-thirds of the home field advantage in MLB comes by virtue of the home plate umpire's bad calls.

We can't expect umpires to be perfect, and in fact, they call strikes and balls correctly 85.6 percent of the time. But the errors they do make don't seem to be random. They favor the home team.

Now that we understand that there is a bias in called balls and strikes, we get a different understanding of why the home team has better hitting and pitching stats. As we've seen, players aren't hitting or throwing any better at home versus on the road. But when you receive more favorable calls at the plate, this directly improves your hitting numbers. There is also an indirect effect. If home batters are benefiting from more favorable pitch calls, they face more favorable pitch counts and are in a better position to swing at pitches to hit. And when home players are put in these situations, it is more likely that their teammates will be on base when they are at the plate, which gives them more opportunities to produce runs. In short, the direct effect from giving home batters fewer strikes and more balls alone seems to account for a sizable fraction of the home team's success in MLB. Add to this the indirect benefits and it could well account for just about all of the home team's advantage.

■ ■ ■

For evidence of official bias in the NFL, it makes sense to start by considering one obvious component in the control of the men in the striped uniforms: penalties. Home teams receive fewer penalties per game than away teams—about half a penalty less per game—and are charged with fewer yards per penalty. Of course, this does not necessarily mean officials are biased. Away teams might commit more violations and play more sloppily or more aggressively. But when we looked at more crucial situations in the NFL—much as with the Leverage Index or the pitch count in baseball—we found that the penalty bias is exaggerated. It turns out that more valuable penalties, those that result in first downs, also favor the home team.

The most compelling evidence of referee influence in the NFL comes from the introduction of instant replay, which gave coaches—and fans, players, and the media—a chance to review and potentially challenge the call on the field. The inauguration of instant-replay challenge came in 1999, and as with the QuesTec results in baseball, it coincided with a decline in the home team success rate in the NFL, from 58.5 percent (from 1985 to 1998) to 56 percent (from 1999 to 2008), a 29.4 percent drop in the home field advantage. Remember, the home advantage starts only when we get above 50 percent.

Coincidence? We can start by looking at turnovers. First, officials wield considerable influence here because they first determine whether there *was* a fumble. Second, they determine which team assumes possession of the football. Before instant replay, home teams enjoyed more than an 8 percent edge in turnovers, losing the ball far less often than road teams. When instant replay came along to challenge wrong calls, the turnover advantage was cut in half.

We can also distinguish between fumbles lost (possession changes hands) and fumbles retained (the team with the ball keeps possession). The home team does not actually fumble or drop the ball less often than the away team—in other words, they aren't "taking care of the ball" any better or worse than the away team. They simply lose fewer fumbles than away teams. After instant replay was installed, however, the home team advantage of *losing* fewer fumbles miraculously disappeared, whereas the frequency of fumbles remained the same. Home teams are as likely as ever to *drop* the ball, but now that visiting teams have the ability to challenge the call, home teams aren't nearly as likely to retain possession.

In close games, when referees' decisions may *really* matter—and when the crowd is really involved—home teams enjoyed a healthy 12 percent advantage in recovering fumbles. After instant replay was installed, that advantage simply vanished.

What about penalties? Instant replay is of limited use to us here because teams can't challenge a penalty call or a noncall. But if we examine the change in penalty discrepancy between home and away teams before and after instant replay, we have a placebo test of sorts. That is, we should not expect to see any changes in penalties. Sure enough, we don't. The discrepancy in number of penalties and yards per penalty given to home versus away teams hardly changed after instant replay. This helps confirm that it is instant replay, not something else, that has driven the recent changes in turnovers and winning percentage of home teams in the NFL.

If referee bias is driving these patterns and instant replay mitigated these biases, we should see that visiting teams are more successful when they challenge a referee's call using instant replay. In other words, if away teams are indeed getting more bad calls than home teams, more of those calls will be overturned on instant replay. We looked at the results of nearly 1,300 instant-replay challenges from 2005 to 2009 to examine the success rate of home team challenges versus away team challenges.

The results? It turns out that away teams are indeed more successful in overturning a call than home teams are, but only modestly so (37 percent versus 35 percent). Both are slightly more successful than official challenges (33 percent), which are challenges initiated by an official in the last two minutes of each half on close plays. These statistics are misleading, though, because as we saw in baseball and soccer, referees are less likely to make biased judgments when the game is no longer in doubt. So what happens if the home team is behind? When the home team is losing, a challenge made by the home team is successful 28.4 percent of the time. But a challenge made by the away team is successful 40.0 percent of the time. Thus, away teams seem to be getting more than their fair share of bad calls when they are winning, which is when bad calls would be most valuable to the home team.

Could referee bias explain a large part of the home field advantage in football? Absolutely. Again we see a dramatic reduction in the home team's edge when instant replay is introduced. Yet instant replay affords each team only a maximum of three incorrect challenges per game and is limited to certain circumstances. Clearly there are other calls not eligible for challenge that could favor the home team, such as penalties. The fact that home teams in football have better offensive stats—such as rushing more successfully and having longer time of possession—could be the result of getting more favorable calls, fewer penalties, and fewer turnovers. If you play at home and sense that you're less likely to get called for a penalty, you may be more inclined to block much more aggressively or challenge a receiver.

■ ■ ■

Recall that in the NBA home and away teams shoot identically from the free throw line. But home teams shoot more free throws than away teams—between 1 and 1.5 more per game. Why? Because away teams are called for more fouls, particularly shooting fouls. Away teams also are called for more turnovers and more violations. These differences could be caused by more aggressive or sloppy play on the part of road teams, which could be more tired because of the lopsided NBA schedule. But they are also con-sistent with referee bias.

To help distinguish sloppy play on the road from referee home bias, let's take a closer look at the *types* of fouls, turnovers, and violations that are committed by home and away teams. Certain fouls, turnovers, and violations require more referee discretion and judgment than others. For example, highly uncertain situations and close calls, where a judgment must be made, allow for greater referee influence, as opposed to something less ambiguous such as a shot clock violation that everyone can easily monitor because the 24-second shot clock is posted above the two baskets and a red light illuminates the glass backboard when the clock expires.

If sloppy or aggressive play by the away team is causing these differences, we should not expect to see the number of violations vary with how ambiguous or uncertain the fouls, turnovers, or vio-lations are regardless of how much referee judgment is required. If you're playing badly, you're probably playing badly across many dimensions of the game.

We looked at calls requiring more or less referee judgment to see whether the home advantage was the same. Loose ball and offensive fouls seem to be the most ambiguous and contentious. Ted Bernhardt, a longtime NBA official, now retired, helped us with our analysis. "Blocking fouls versus charging fouls are by far the hardest calls to make," he says. It turns out that offensive and loose ball fouls go the home team's way at twice the rate of other personal fouls. We can also look at fouls that are more valuable, such as those that cause a change of possession. These fouls are al-most *four* times more likely to go the home team's way than fouls that don't cause a change of possession.

What about turnovers and violations? Turnovers from shot clock violations, which aren't particularly ambiguous or contro-versial, are no different for home or away teams. Turnovers from five-second violations on inbounds plays, which are also fairly unambiguous because everyone can count (though referees may count a little slower or faster than everyone else and there is no clock indicating when five seconds has elapsed), are also not very different for home and away teams (in fact, home teams receive slightly more five-second violations).

If, however, we look at the most ambiguous turnover calls requiring the most judgment, such as palming and traveling, we see huge differences in home and away numbers. The chance of a visiting player getting called for traveling is 15 percent higher than it is for a home team player. The fact that ambiguous fouls and turnovers tend to go the home team's way and unambiguous ones don't is hard to reconcile with sloppy play on the part of visiting teams. But it's exactly what you would expect from referee bias.

Identifying refereeing bias in the NBA is especially hard because context is so important, and some of the most controversial "calls" in basketball are in fact "no calls"—when a call is not made. But the evidence seems to suggest ref bias toward the home team. If bias clearly exists in soccer, baseball, and football, isn't it reasonable to suspect that NBA referees are vulnerable to the same influences?

Remember the Portland Trail Blazers playing so sluggishly in that dreary midweek road loss to the San Antonio Spurs? On the road in San Antonio, the Trail Blazers committed 13 fouls, the Spurs 14; each team had six turnovers. But if we look at the *types* of fouls and turnovers over which referees have more influence, we see that the Blazers were whistled for twice as many loose ball fouls as the Spurs and that five of the Blazers' six turnovers were on judgment calls made by the referee (one traveling, two ambiguous lost balls out of bounds, one offensive goaltending, and one questionable kicked ball). By contrast, all six Spurs turnovers were unambiguous (five bad passes/steals and one shot clock violation). In addition, more of the calls against the Blazers resulted in a change of possession favoring the Spurs. Perhaps it's not so surprising that the Spurs won.

Recall that only a few nights later the teams met again, this time in Portland. The Blazers won by 18 points. The same advantages conferred on the home team were present, though this time it was the Trail Blazers who were the beneficiaries. The Spurs were whistled for 25 fouls and 16 turnovers, compared with Portland, which had 18 fouls and 13 turnovers. The types of foul calls and turnovers tell an even stronger story. Among the visiting Spurs' 16 turnovers, 11 were of the more ambiguous variety, including a couple of debatable lost balls out of bounds, and two Spurs players were even called for palming. (To give you an idea of how rarely palming is called in the NBA, on average there is one palming call every five or six games.) Of the Blazers' 13 turnovers, 10 were unambiguous, consisting of two shot clock violations and bad passes that were stolen or thrown out of bounds. There were ten situations in which the ball was tipped out of bounds—eight went the Trail Blazers' way. More fouls resulting in a change of possession went Portland's way as well. If we tally the numbers across the two games, ambiguous turnovers went the home team's way 85 percent of the time, ambiguous fouls were charged to the visiting team 72 percent of the time, and the home team won by a collective 33-point margin.

So how much of the home court advantage in the NBA is due to referee bias? If we attribute the differences in free throw attempts to referee bias, this would account for 0.8 points per game. That alone accounts for almost one-fourth of the NBA home court advantage of 3.4 points per game. If we gave credit to the referees for the more ambiguous turnover differences and computed the value of those turnovers, this would also capture another quarter of the home team's advantage. Attributing some of the other foul differences to the referees and adding the effects of those fouls (other than free throws) on the game, this brings the total to about three-quarters of the home team's advantage. And remember, scheduling in the NBA explained about 21 percent of the home team's success, as well. That adds up to nearly all of the NBA home court advantage.

Long story short, referee bias could well be the *main reason* for home court advantage in basketball. And if the refs call turnovers and fouls in the home team's favor, we can assume they make other biased calls in favor of the home team that we cannot see or measure.

■ ■ ■

What about the NHL? By now you can probably guess what we found. Home teams in hockey get 20 percent fewer penalties called on them and receive fewer minutes in the box per penalty. (In other words, home teams are not just penalized less often but penalized for less severe violations.) The net result is that on average per game, home teams get two and a half more minutes of power play opportunities—a one-man advantage during which goals often are scored—than away teams. That is a *huge* advantage. To provide some perspective, the average NHL team succeeds in scoring a goal during a two-minute power play about 20 percent of the time. So if you take the power play advantage and multiply it by the 20 percent success rate (per two minutes), this gives the home team a 0.25-goal advantage per game. The average point differential between home and away teams in the NHL is 0.30 goals per game, so this alone accounts for more than 80 percent of the home ice advantage in hockey.

But is the penalty difference driven by refereeing bias? Repeating the same exercise we conducted for the NBA, we looked at more ambiguous calls—holding, hooking, cross-checking, boarding, tripping—and found that these penalties in particular went the home team's way. Less ambiguous calls such as too many men on the ice, illegal equipment, delay of game from sending the puck into the stands, and fighting had much less home team bias. Again, this is consistent with officiating bias—and not with tired or sloppy play from visiting teams.

Also, don't forget the shootout results we discussed earlier. Remember, in a shootout we found no home ice advantage. Not coincidentally, this is the only part of the game in which the referee essentially plays no role.

■ ■ ■

The fact that we can identify an officiating bias toward the home team is unsettling—that this may be the chief reason home field advantage exists in every sport is *very* unsettling. But *why* are officials biased toward the home team?

WHY DO OFFICIALS FAVOR THE HOME TEAM?

First let's be clear: Is there a conspiracy afoot in which officials are somehow *instructed* to rule in favor of the home team, especially since the league has an economic incentive to boost home team wins? Almost unquestionably no. We're convinced that the vast majority of, if not all, officials are upstanding professionals, uncorrupted and incorruptible, consciously doing their best to ensure fairness. All things considered, they do a remarkable job.

They are not, however, immune to human psychology, and that's where we think the explanation for home team bias resides. Despite fans' claims to the contrary, referees are, finally, human. Psychology finds that social influence is a powerful force that can affect human behavior and decisions *without the subjects even being aware of it*. Psychologists call this influence conformity because it causes the subject's opinion to conform to a group's opinion. This influence can come from social pressure or from an ambiguous situation in which someone seeks information from a group.

In 1935, the psychologist Muzafer Sherif conducted a study about conformity, using a small point of light in an otherwise dark or featureless environment. Because of the way the human eye works, the light appears to move, but the amplitude of the movements is undefined—individual observers set their own frames of reference to judge amplitude and direction. Therefore, each individual saw the "movement" differently and to differing degrees.

When participants were asked individually to estimate how far the light had moved, as one would expect, they gave widely varying answers. Then they were retested in groups of three. The composition of the group was manipulated; testers put together two people whose estimate of the light movement when alone was very similar and one person whose estimate was very different. Each person in the group had to say aloud how far he or she thought the light had moved. Sherif found that over numerous trials, the group converged on a common estimate. The subject whose estimate of movement had been vastly different from that of the other two in the group came to conform to the majority view.

More important, when interviewed afterward, the subject whose initial estimate had been very different now *believed* his or her initial estimate was wrong. That is, that subject did not succumb to social pressure and state something he or she didn't believe; his or her actual perception of the light's movement had changed. The experiment demonstrated that when placed in an ambiguous situation, a person will look to others for guidance or additional information to help make the "right" decision.

After the Sherif study, Solomon Asch, a pioneer of social psychology, conducted an experiment in which he asked participants to look at two cards and decide which line (A, B, or C) on the card on the right in the following illustration was most like the line on the card on the left.

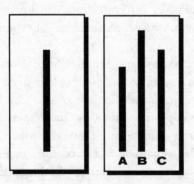

The answer, you probably guessed, is C. The participants, though, were asked to make this assessment in a group setting. Asch had put one unwitting subject in a room with seven confederates, or actors. The actors were told in advance how to respond.

Each person in the room gave his or her answer, and the "real" participant offered his or her answer second to last. In most of the cases, the subject yielded to the majority at least once, even though he or she suspected it was wrong.

Asked why they readily conformed to the group even though they felt the answer was wrong, most participants said that they did not really believe their answer; rather, they went along with the others for fear of being ridiculed or thought "peculiar." A few, however, said that they really did believe the group's answers were correct. Asch also found that subjects felt enormous stress when making these decisions; giving a response that was at odds with the majority caused anxiety, even though they knew they were right.

The takeaway here is that human beings conform for two reasons: (1) because they want to fit in with the group and (2) because they believe the group is better informed than they are. Makes sense, right? If you are asked to make a decision and are unsure of your answer, wouldn't you look for other cues and signals to improve that answer? And don't you accord weight to people's answers by the confidence with which they provide them? After a difficult test in school, who hasn't polled other classmates for the answer to a question, paying particular attention to the responses of the known "A" students?

■ ■ ■

Now, back to referees. When humans are faced with enormous pressure—say, making a crucial call with a rabid crowd yelling, taunting, and chanting a few feet away—it is natural to want to alleviate that pressure. By making snap-judgment calls in favor of the home team, referees, whether they consciously appreciate it or not, are relieving some of that stress. They may also be taking a cue from the crowd when trying to make the right call, especially in an uncertain situation. They're not sure whether that tailing 95-mph fastball crossed the strike zone, but again, even if it's subconsciously, the crowd's reaction may provide a useful signal that changes their perception.

If beliefs are being changed by the environment, as psychology shows, referees aren't necessarily consciously favoring the home team but are doing what they believe is right. It's just that their perceptions have been altered. In trying to make the right call, they are conforming to a larger group's opinion, swayed by tens of thousands of people witnessing the exact same play they did. As the saying goes in psychology, "I'll see it when I believe it." Referees, it's safe to assume, do not intend this favoritism. They're probably not even aware of it. But it is a natural human response.

Remember, too, that on top of the anxiety caused by passionate and sometimes angry fans, the refs receive stress from their supervisors and superiors. In a variety of ways—some subtle, some not—officials must take in cues that the league has an economic incentive for home teams to do well. If your boss sent a subtle but unmistakable message that Outcome A was preferable to Outcome B, when you were forced to make a difficult, uncertain, and quick decision, how would you be inclined to act?

Let's look at our previous results on referees through the lens of psychology and our understanding of the human propensity to conform. The extra injury time in soccer? It is probably a response to social pressure, that is, the desire to please the crowd—and in some cases preserve personal safety. The strike-ball discrepancy in baseball and similar disparities in fouls and turnovers in basketball, along with penalties and turnovers in football and hockey, may also be the result of "informational conformity" in the face of social pressure—using the crowd as a cue to resolve an uncertain or ambiguous situation.

If this is true, psychology suggests that both the crowd size and the uncertainty or ambiguity of the situation should make a difference. Home team favoritism therefore should be greater the larger and more relevant the crowd and the more ambiguous the situation. We've already shown in a variety of ways how the more ambiguous the call—whether it is a 90-mph pitch on the corner of the strike zone in baseball, a fumbled football, a two- or three-step move without dribbling in basketball, or a questionable check in hockey—the more severe the home advantage.

What about the size of the crowd? Recall the original study of the Spanish La Liga. The authors found that the bias in regard to extra time was even more evident when the crowd was larger. Similarly, the studies in the English Premier League, Italian Serie A, German Bundesliga, and MLS also found that referee favoritism was more apparent when attendance was higher. Maybe most interesting was the study conducted in Germany, where many of the soccer stadiums also house a running track that acts as a moat, separating the stands from the field of play. In those stadiums, the referees are more removed from the fans. Guess what? The bias referees usually exhibit for the home team gets cut in half in those stadiums but is the same as it is in other leagues for German stadiums that do not contain a track. In the three European soccer leagues we examined, attendance also had a marked effect on the number of red and yellow cards the visiting team received relative to the home team. Other studies have also linked attendance to penalties and fouls, showing that the bias in favor of the home team grows with the crowd.

What about the extra walks awarded to home teams and the extra strikeouts imposed on away teams by the home plate umpire? These, too, occur predominantly in high-attendance games and are not present in the games with the lowest attendance. The chart below shows the net strikeout and walk advantage to home teams from bad umpire calls, reported separately for the games with the lowest and highest attendance (bottom and top fifth of attended games). Although there is virtually no home team strikeout or walk advantage in the least-attended games, the highest fifth of attended games account for more than half of the entire strikeout and walk advantage given to home teams each season. In the highest-attended games, home teams are given 263 fewer strikeouts than their opponents. In the lowest-attended games, that falls to 33 fewer strikeouts. Similarly, the home team receives 93 more walks than the visitors from bad umpire calls in the most-attended games relative to the least-attended ones.

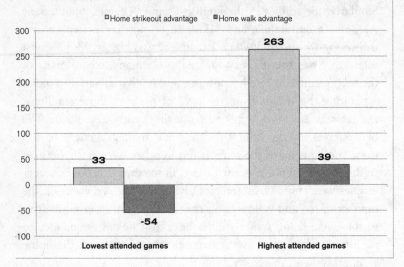

HOME TEAM ADVANTAGE IN STRIKEOUTS AND WALKS FROM UMPIRE INCORRECT CALLS IN LOW AND HIGH ATTENDANCE GAMES

□ Home strikeout advantage ■ Home walk advantage

Lowest attended games: 33, -54
Highest attended games: 263, 39

In the NBA, crowd size also affects the home-away differences, particularly for the more ambiguous calls. Recall how traveling is called 15 percent less often against home players. Looking at NBA games in the bottom fifth of attendance, this discrepancy goes down to 6 percent. But if we look at the most-attended games, the home team is 28 percent less likely to be called for traveling.

In the NHL, the bigger the crowd, the more penalties, fouls, and close calls that go against the visiting team, and once again, the effects are greatest for more ambiguous calls. Even in the NFL, in which most games are sold out, the home-away discrepancies in penalties and turnovers increase with crowd size. With virtually every discretionary official's call—in virtually every sport—the home advantage is significantly larger when the crowd is bigger.

In fact, in the least-attended games in each sport, the home field advantage all but vanishes. In MLB, if you look at the 20 percent least-attended games, the home field advantage is only 50.7

percent. In other words, home and away teams are about equally likely to win when the crowd is small. In the one-fifth of games with the highest attendance, however, home teams win 55 percent of the time in MLB. In the NBA, the least-attended games are won by the home team only 55 percent of the time, and the most attended games 69 percent of the time. In the NHL, the home team wins only 52 percent of the time in the lowest-attended games but 60 percent of the time in the highest-attended games. And in European soccer, the home team wins 57 percent of the time in the lowest-attended games and an astonishing 78 percent of the time in the highest-attended matches.

Wait a second, you might say. Doesn't this stand to reason? After all, crappy teams draw crappy crowds, so the games with the empty seats usually involve the worst teams. Never mind official bias; just look at the standings. You'd expect the Pittsburgh Pirates or the New Jersey Nets—lame teams, lame crowds—to win fewer home games than, say, the Boston Red Sox or the Los Angeles Lakers. True, but even after adjusting for the strength of the team we find similar effects. Also, it doesn't matter as much as you might think, because when a bad team hosts a good team, attendance often spikes. When LeBron James and the Miami Heat visit Memphis or Milwaukee, the crowds swell. The worst-attended games usually involve two terrible teams, and the most-attended games feature two great teams. So it turns out there isn't much of a difference in ability between the two teams in either case.

Still not convinced by the psychological explanation for referee bias? Consider a final study, this one performed in 2001. Researchers recorded videos of soccer matches, focusing on tackles during the game, and showed them to two groups of referees. The first group was shown the tackles with the crowd noise audible. The second group was shown the same tackles with the crowd noise muted. Both sets of referees were asked to make calls on the tackles they saw. The referees who watched the tackles with the crowd noise audible were much more likely to call the tackles *with* the crowd. That is, tackles made against the home team (where the crowd complained loudly) were more likely to be called fouls and tackles made by the home team were less likely to be called fouls. The referees who viewed the tackles in silence showed no bias.

You probably guessed correctly which group of referees made calls consistent with the actual calls made on the field. Yes, the ones who could hear the crowd noise. Not only that, but the referees watching with sound also reported more anxiety and uncertainty regarding their calls, consistent with the stress they felt from the crowd. Imagine how much more intense that stress would have been if they had been on the actual field of play.

But perhaps the most convincing evidence for the effect of crowds on referees occurred when *no fans* were present. On February 2, 2007, supporters of two soccer clubs in Italy—Calcio Catania and Palermo Calcio—clashed with each other and police. It was a typical hooligan-induced riot, and following the episode the Italian government forced teams with deficient security standards at their stadiums to play their home games without *any* spectators present. Two economists (and soccer fanatics) from Sweden, Per Pettersson-Lidbom and Michael Priks, collected the data from the 21 soccer matches that were played before empty bleachers.

What they found was amazing. When home teams played without spectators, the normal foul rate, yellow card, and red card advantage afforded home teams disappeared entirely. Looking at the same team with the same crew of officials, the authors found that when spectators were no longer present, the home bias in favorable calls dropped by 23 to 70 percent, depending on the type of calls (a decline of 23 percent for fouls, 26 percent for yellow cards, and 70 percent for red cards). That is, the *same* referee overseeing the *same* two teams in the *same* stadium behaved dramatically differently when spectators were present versus when no one was watching.

When the economists also looked at player behavior, they found that, unlike the referees, the players did *not* seem to play any differently when the crowd was there yelling versus in an empty, silent stadium. Home and away players shot the same percentage of goals on target, passed with the same accuracy, and had the same number of tackles as they normally do. The absence of the crowd did not seem to have any effect on their performance. This is in keeping with what we saw for NBA foul shooters, hockey penalty shots, and MLB batters and pitchers: Crowds don't appear to have much effect on athletes.

So it is that we assert that referee bias from social influence not only is present but is *the leading cause of the home field advantage.*

■ ■ ■

We started with three questions that any explanation of the home field advantage must address: (1) Why does it differ across sports? (2) Why is it the same for a particular sport no matter where the game is played? (3) Why hasn't it changed much over time?

To answer the first question, if the refs are responsible for the home advantage, it must be the case that referees are more important or have more influence in some sports (say, soccer, in which home teams have the greatest success) than in others (such as baseball, where the advantage is weakest). As it turns out, this is precisely the case. In soccer, the official has an enormous influence on the outcome of the game. One additional penalty, free kick, or foul can easily decide a game, in which one goal is often all that separates the two teams. In basketball, which has the second highest home team advantage, the official could call a foul on almost every play. By contrast, the umpire's role in baseball is limited relative to other sports. Most plays and most calls are fairly unambiguous; a home run is a home run—either it cleared the fence or it didn't. Most force-outs are not close. Sure, the umpire has discretion over called balls and strikes, but more than half the time the batter swings, eliminating umpire judgment.

In addition, crowd size, which we contend affects referee judgment, has more influence in the sports with the greatest home field advantage. Crowd size matters most in soccer (the sport with the highest home field advantage) and least in baseball (the sport with the lowest home team winning percentage) and is somewhere in between for the other sports. This is also consistent with referees mattering more in some sports (soccer) than others (baseball).

To answer the second question, referee bias also explains why the home field advantage is the same for a particular sport no matter where it is played. Whether baseball is being played in the United States or Japan, whether it's basketball in the NBA, WNBA, or NCAA or soccer in France versus South Africa, the rules and, more important, the role of the referee are essentially the same, no matter where the game is played.

Finally, to answer the third question, referee bias also explains why the home team's success rate hasn't changed over a century. Although sports have altered their rules over time—raising and lowering the pitcher's mound, introducing a shot clock and the three-point line—the official's role in the game hasn't changed much. Umpires still call balls and strikes, referees still call fouls and penalties, and for over a century these calls have been made by human beings—none of them immune from human psychology.

Although we will never be able to measure or test all the decisions an official makes, if we can see that some biased judgments are being made, it is likely there are other biases going the home team's way that we don't see. Think of the father who comes home early from work and catches his teenage daughter kissing her boyfriend. He's upset about the kiss, but he's more upset about what else she might be doing when he doesn't happen to be looking.

■ ■ ■

Knowing what we now know, let's revisit that Cubs-Brewers game, the ten-inning affair that ruined a summer day for Jack Moore of Trempealeau, Wisconsin. You wouldn't deduce this by scanning a conventional box score or watching a *SportsCenter* highlight. But after revving up the Pitch f/x results, it becomes clear that when the umpire erroneously called a ball on a 3–2 pitch in the bottom of the tenth inning, enabling the winning run to score for the hosting Cubs, it marked the culmination of an afternoon filled with unfavorable decisions against the visiting Brewers.

According to Pitch f/x, Cubs hitters failed to swing at 25 pitches that were strikes. However, nearly a third of them were incorrectly called balls. As for the Brewers, they failed to swing at 32 pitches in the strike zone, only a quarter of which were called incorrectly as balls. Advantage, Cubs. In high-leverage situations, when batters had three balls, *not a single* strike was called on a Cubs hitter even when the ball was in the strike zone—including, of course, the final pitch of the game. But for Brewers hitters facing three-ball counts, *every* pitch in the strike zone was called a strike and *half* the pitches outside the strike zone were called strikes! Big advantage, Cubs.

Overall, the Brewers were deprived of three walks to which they were entitled and the Cubs were given two walks on strikes that were erroneously called balls, including the game-winner. That's a difference of five base runners in a game that ended with a final score of 2–1 in extra innings.

One last point: Recall how the closer officials are to the crowds, the more likely they are to favor the home team. Wrigley Field has about the smallest amount of foul territory in the Major Leagues, so the umpire is uncommonly close to the restless natives. And remember how attendance influences the home field advantage. Wrigley Field seats 41,118 fans and is generally nearly full. In fact, despite a long history of losing seasons, the Cubs have won 54 percent of their home games—above the league average. (It's just that they have been terrible on the road.) That particular afternoon drew a crowd of 41,204—more than 100 percent of capacity with standing room only.

When that mass of humanity on Chicago's North Side yelled at the players, they weren't affecting the outcome. When they yelled at the umpire, well . . . that's another story entirely.

THERE'S NO *I* IN *TEAM*

But there is an *m* and an *e*

After concluding that defense *doesn't* necessarily win champion-ships, we decided to examine another shopworn bit of sports wis-dom. Before young athletes are capable of lacing their sneakers and putting on their cleats, they're invariably taught, "There's no *I* in *team*." This spelling lesson is, of course, meant to reinforce the virtues of teamwork, stressing the importance of unity and the corrosive effects of attempts at personal glory. But does it ac-curately capture reality?

It's in basketball that the no-I-in-team cliché is most often tossed around. If we were to compile a list of the top, say, five or six NBA players over the last 20 years, it probably would include Michael Jordan, Kobe Bryant, Tim Duncan, Shaquille O'Neal, Hakeem Olajuwon, and LeBron James. If there were no *I* in *team*, those stars wouldn't much matter. A team that formed a sym-phonic whole, with five players suppressing and sublimating ego for a common goal, could surmount teams with the "I" players, stars willing and able to play selfishly when the situation calls for it. But that's rarely the case. Since 1991, every year at least one of those players has appeared in the NBA finals. Go back another de-cade and add Larry Bird and Magic Johnson and now at least one of the top eight players has been featured in all but one NBA finals series for the last 30 years. In other words, a remarkably small number of select players have led their teams to the vast majority of NBA titles. Lacking one of the best players in history has all but precluded a team from winning.

We wondered how likely it is that an NBA team without a superstar wins a championship, makes it to the finals, or even makes it to the playoffs. We can define superstars in various ways, such as first-team all-stars, top five MVP vote-getters, or even those with the top five salaries. Pick your definition; it doesn't much matter. The chart below shows what we found for the NBA.

A team with no starting all-star on the roster has virtually no chance—precisely, it's 0.9 percent—of winning the NBA championship. More than 85 percent of NBA finals involve a superstar player and more than 90 percent of NBA titles belong to a team with a superstar. The graph also shows, not surprisingly, that as a team gains superstars, its chances of winning a title improve

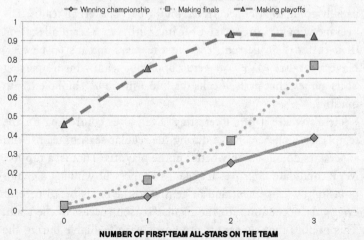

PROBABILITY OF WINNING AN NBA CHAMPIONSHIP, MAKING FINALS, OR MAKING PLAYOFFS GIVEN NUMBER OF SUPERSTARS ON THE TEAM

◆ Winning championship ▪ Making finals ▲ Making playoffs

NUMBER OF FIRST-TEAM ALL-STARS ON THE TEAM

dramatically. One first-team all-star on the roster yields a 7.1 percent chance of winning a championship and a 16 percent chance of making it to the finals. A team fortunate enough to have two first-team all-star players stands a 25 percent chance of winning a championship and a 37 percent chance of making the finals. On the rare occasions when a team was somehow able to attract *three* first-team all-stars, it won a championship 39 percent of the time and made the finals 77 percent of the time.

The numbers are even more striking when we consider the top five MVP vote recipients. A team with one of those players stands a 15 percent chance of winning it all and a 31 percent chance of making the finals. Having two of those players yields a 48 percent chance of winning the championship and a 70 percent chance of making the finals.

Really, it's no mystery why the Miami Heat fans celebrated deliriously when the team lured LeBron James in the summer of 2010. When James "took his talents to South Beach" and joined Dwyane Wade and Chris Bosh, it made the Heat a virtual lock to go deep into the playoffs. Having the MVP (James) in addition to two other all-stars makes the Heat 98 percent likely to make the playoffs, 70 percent likely to make the finals, and 36 percent likely to win it all.

At some level this stands to reason, right? The superstars are usually going to be concentrated on the best basketball teams. The average winning percentage of teams with a first-team all-star on their roster is 56 percent. Two first-teamers and it's 63 percent. Have a top-five MVP vote-getter and your team wins 64 percent of its games; that in itself calls the "no-*I*-in-*team*" shibboleth into question.

Here's where it gets interesting. Even after controlling for the team's winning percentage during the regular season, teams with superstars do measurably better in the playoffs. That is, a top-five MVP candidate improves his team's chances of winning a championship by 12 percent and of getting to the finals by 23 percent even *after* accounting for the regular season success of the team. This implies that superstars are particularly valuable during the playoffs—ironically, the time when "team" is relentlessly stressed by coaches, media, and analysts.

What about the notion that a lineup of five solid players is better than a starting five of one superstar and four serviceable supporting role players? One way to test this idea is to look at the disparity among a team's starters in terms of talent. Controlling for the same level of ability, do basketball teams with more evenly distributed talent fare better than teams with more dispersed talent? Measuring talent is difficult, but one reasonable metric is salary. Controlling for the average salary and winning percentage of teams, do teams with bigger *differences* in salaries among their starting players fare worse than teams whose salaries are spread more evenly among the players?

We find the opposite, in fact. Teams with more variable talent across their players are more likely to make the finals and more likely to win a championship than teams with more uniformly distributed talent. Again, this suggests that a superstar with a relatively weak supporting cast fares better than the team with five good players.

The same holds for the NHL and even soccer. Without a prolific goal scorer and/or goaltender/goalkeeper, survival in the postseason tends to be short-lived. There may be an *I* in Major League Baseball, too, but the effects are considerably weaker, in part because the team is larger, making it harder for one player to change the overall make-up of the team. Nonetheless, the bulk of World Series titles and appearances belong to the teams with a handful of elite superstars, both hitters and pitchers. But there are some examples of championship teams without a starting all-star player. (Name a "star" on the 2003 Florida Marlins. How about the 2002 Angels?) But this makes sense. Although we focus on individual achievements in baseball, it's hard for any one player other than the starting pitcher, who pitches only one game out of five, to take control of a game. Even the best hitters come to bat only once every nine times.

▪ ▪ ▪

If the evidence suggests that in basketball, hockey, and soccer a handful of individual players are extremely valuable for success, why do so many coaches and commentators place such heavy emphasis on the team? Perhaps it's because admitting that in these sports the star matters as much as he does blunts the incentive for the rest of the team. Though teammates may be less valuable than the stars, they still have *some* value. They're needed to grab rebounds, pass, block, chase loose balls, and defend. Sure, the superstar makes a big difference, but he can't do it alone. In that sense, the team certainly does matter.

Stars tend to recognize this delicate balance. Remember how lustily Michael Jordan embraced the conventional wisdom that "defense wins championships," a phrase that galvanized his teammates? Nonetheless even Jordan felt differently about the "no-*I*-in-*team*" truism. He recognized that not all players were created equal.

At his 2009 induction into the Basketball Hall of Fame, Jordan gave a speech that revealed much about his turbo-powered competitive drive. He told a story of once scoring 20 consecutive points late in a game to lead the Chicago Bulls to victory. Afterward, he was admonished by Tex Winter, the Bulls' eminent longtime assistant coach, "Michael, there's no *I* in *team*." Jordan recalled his response: "I looked back at Tex, and said, 'There's an *I* in *win*. So which way do you want it?'"

OFF THE CHART

How Mike McCoy came to dominate the NFL draft

At the Dallas Cowboys' team headquarters in spring 1991, a sense of optimism was leavened with a sense of unease. The most iconic franchise in the NFL had recently been sold to a swashbuckler from Arkansas, Jerry Jones. Just as Jones had made a fortune in the oil and gas business by taking bold risks, he'd leveraged his entire net worth to buy the 'Boys in 1989 for $140 million—$65 million for the team and $75 million for the stadium. Jones paid $90 million in cash and borrowed the rest against personal assets. His interest payments were $40,000 a day. His banker told him he was nuts. His father told him the same thing, but hey, this was the Dallas Cowboys.

The beginning of the Jones regime did not portend greatness. One of his first acts was to fire the longtime coach, the legendary and dignified Tom Landry, and replace him with his polar opposite, Jimmy Johnson, a brash (if notably well-coiffed) renegade who'd never coached in the NFL. He was fresh from a spectacularly successful, spectacularly controversial tenure at the University of Miami. Johnson was an old teammate and running buddy of Jones's from their days at the University of Arkansas, and that counted for plenty. Jones conferred full football decision-making powers on Johnson, forcing out Tex Schramm, the only president the Cowboys had ever employed and, as his name suggests, a man with deep roots in the state. Worst of all, the Cowboys put a rotten product on the field, winning just 8 of 32 games in Jones's first two seasons as owner.

As is so often the case for sad-sack teams in the socialist world of team sports, hope for the Cowboys came in the form of draft picks. The best picks go to the worst teams, though this can be fool's gold. In 1991 Dallas held a Texas-sized helping of selections: ten picks in the first four rounds, five in the first round alone, including the very first pick. When Johnson crowed, "We're dictatin' this whole draft," he wasn't met with much resistance. But if this draft had the potential to be a turning point for the franchise, it also had the potential to be Jones's personal Waterloo: You picked early and often and you still couldn't turn the franchise around?

Reflecting the new owner's passion for speculation, Dallas traded more than any other team in the league. In the 26 months since Jones had bought the team, the Cowboys had made 29 deals, including the infamous "Herschel Walker trade" with the Minnesota Vikings, an 18-player swap that was less a transaction than an act of larceny. In exchange for the aging Walker and four modest picks, the Cowboys received a bounty of five players plus eight future picks. Two decades later, this still stands as a benchmark for lopsided trades. It ended up paying substantial dividends for the Cowboys in the years to come, but Johnson recalls that at the time it made other teams wary of doing business with the Cowboys for fear of getting similarly scalped.

However, with their stockpile of selections, it was logical to assume that on draft day other teams would offer to trade picks. That created a problem: With the clock ticking, how was Johnson to know whether he'd be better served to entertain the offer of swapping, say, one of the team's third-round picks for Green Bay's fifth- and seventh-round picks? "I can't assess value that fast," Johnson complained at a predraft meeting among Dallas executives. "No one can!" There were nods all around.

Mike McCoy, a Cowboys executive and minority owner, piped up: "Let me see what I can come up with to fix that."

In 1981, McCoy, then an Arkansas petroleum engineer, had partnered with Jones to form the Arkoma Production Company. Jones made the deals, and McCoy was the driller. As *Inc.* magazine once put it, they perfected a low-risk strategy of drilling holes that other companies suspected were fertile but wouldn't commit to exploring. The success rate on "wildcats"—speculative wells that come cheap but seldom yield a payoff—is around 5 percent. According to *Inc.*, Jones and McCoy struck 2,000 wells and made money on more than 500 of them. In 1986, a state-run gas company in Arkansas, Arkla, bought Arkoma for $175 million. (Aside: Sheffield Nelson, the former head of Arkla, ran for governor in 1990 and lost to the incumbent—an ambitious Democrat named Bill Clinton—in part because of questions surrounding the generous payout he authorized to Jones and McCoy.) When Jones purchased the Cowboys, McCoy joined him. Jones, after all, often referred to McCoy as "one of the brightest minds I've ever been around."

This business of trying to quantify draft picks? As McCoy saw it, he'd spent his entire life solving numerical puzzles and trying to tilt the odds in his favor. This was simply another application. He recalls thinking to himself: "How hard could this really be?"

Not hard at all, it turned out. McCoy asked Dallas's player personnel department for a list of all the NFL trades that had been conducted on draft day going back four years. He assigned an arbitrary point value to the first pick in each draft round and then used all the prior draft trades to refine the relative values of every pick. As McCoy recalls, "The point was to create a graphical depiction of how the NFL valued draft picks, based upon their own actions—not how they should have been valued."

McCoy didn't make any subjective judgments; he simply took the existing information and plotted the data points. After two days of "fiddling" (his word) and plotting picks on a graph, he presented a chart that assigned a numerical value to every draft position. "It was basically a price list," McCoy says. "It was what

Walmart would do, only with football players, not jeans or tooth-paste." The first pick of the first round was worth X. The last pick of the last round—known as Mr. Irrelevant—was worth only Y. A sample of the point totals appears in the table below.

A SAMPLE OF THE POINT VALUES
FROM MCCOY'S CHART

First pick, round one:	3000
Second pick, round one:	2600
Tenth pick, round one:	1300
Thirty-second pick, round one:	590
First pick, round two:	580
First pick, round three:	265
First pick, round four:	112
Final pick:	0.4

According to the chart, the value of the first pick in the draft (3000) was equal to the combined value of the sixth pick (1600) and eighth pick (1400) but more than that of the final four picks of the first round (640 + 620 + 600 + 590 = 2450) combined.

McCoy is quick to admit that it was a crass calculation, hardly built on rigorous econometrics. But armed with the chart, the Cowboys approached draft day in 1991 with supreme confidence. Whenever a trade offer came over the transom, they'd simply consult their conversion chart, make a few calculations, and determine whether it was worthwhile. If it was "below the line," it was best to pass; above the line, it was probably a steal. After using the first pick of the first round to select Russell Maryland, a defensive tackle Johnson had once recruited to the University of Miami, Dallas traded two of its first-round selections. They wheeled, they dealed, they took 17 players in all (including three eventual Pro Bowl players); by the time the weekend was over, the Cowboys' inner sanctum was drunk with confidence. Draining a few bottles of beer, Johnson told an embedded *Sports Illustrated* reporter, "We'll be good; big-time good. There's no doubt in anybody's mind here. . . . I couldn't care less what the people out there think of us."

Sure enough, in the ensuing years, the Cowboys gained back the aura of America's Team. The Blue and Silver mystique returned as Dallas won three Super Bowls over the next five years.

The franchise's turnaround was due in no small part to the Cowboys' exceptional success on draft day. In the five years after the unveiling of the chart, Dallas selected 15 starters and five Pro Bowl players. "It got to the point," says McCoy, "where teams were afraid to trade with us." Jerry Jones soon began referring to the chart as Dallas's secret weapon.

There were ancillary benefits as well. Using McCoy's bible, the Cowboys were able to identify other teams that consistently overpaid for talent. "Those were the teams we wanted to call!" says McCoy. In 1999, for instance, the New Orleans Saints famously traded eight draft picks, including all their 1999 selections, to the Washington Redskins in order to draft Ricky Williams with the fifth pick in the first round. At least according to the values of the chart, New Orleans had overpaid to comical proportions, and this did not go unnoticed in the Cowboys' war room.* Note to self: *Trade with New Orleans whenever possible.*

It was probably inevitable, but the Cowboys' secret weapon didn't stay secret forever. With the Cowboys winning so prodigiously, it was only natural that their coaches and coordinators would attract the interest of other teams. Before Dave Wannstedt went to coach the Chicago Bears or Norv Turner took the head job with the Washington Redskins, they made sure to grab a copy of the franchise's sacred text as they packed. Dallas scouts and front office employees also took the chart with them as they decamped for other teams. Within a decade, most, if not all, teams in the league had a purloined copy of McCoy's creation.

---

* Ironically, Washington used those eight picks to remarkably bad effect, selecting bust after bust. Williams, meanwhile, led by the rapper Master P, negotiated perhaps the most lampooned contract in sports history, a deal laden with performance-based incentives, few of which Williams managed to meet, thereby sparing the Saints millions. So what should have been a disastrous trade for New Orleans was more or less a wash, equally bad for both parties.

In 1996, Jones bought out his buddy McCoy, though to this day the two remain close friends and business partners in natural gas ventures. By then the Cowboys were worth $300 million, more than double Jones's purchase price. (Today, Wall Street values the franchise at close to $2 billion—more than 12 times what Jones paid.) Now an investor in Dallas, McCoy chuckles when he considers the legacy of his creation. "I guess it leveled the playing field and made trading easier because everyone could point to the chart and cover their butt," he said. Then he added forlornly, "But after a while, you couldn't [fleece] other teams the way we used to."

Or could you? After all, McCoy's creation was an artifact of what teams *did,* not necessarily what they *should do.* The chart provided the average value of draft picks based on actual trades teams made, and so the Cowboys could tell whether a certain trade was above or below the average value other NFL teams placed on those players. But what if the average value teams placed on draft picks was wrong? Sure, every team now had a copy of the chart, but few teams double-checked McCoy's valuations or updated them in accordance with salary cap changes or, more important, *the performance of the actual picks.* Did anyone stop to check whether the number one pick *really was* more than twice as good as the number eight pick, as the chart dictated? No. "We're football guys, not math majors," said one executive. "We're all using this document that a buddy of Jerry Jones put together using picks in, like, the late eighties? Now that you put it like that, it's probably not so smart."

Definitely not so smart, at least according to two prominent behavioral economists who studied the NFL draft. In 2004, Richard Thaler, a professor of behavioral economics at the University of Chicago, and Cade Massey at Yale were watching the NFL draft. With the first pick, the San Diego Chargers chose quarterback Eli Manning, the brother of perhaps the best quarterback in the league, Peyton Manning, and the son of longtime NFL quarterback Archie Manning. The New York Giants held the number four pick and were in the market for a premier quarterback as well. It was no secret they coveted Manning and thought he was the best prospect.

As the estimable Peter King from *Sports Illustrated* reported at the time, during the 15 minutes the Giants had to make their selection, they were ambushed with two very different options. Option 1 was to make a trade with San Diego in which the Giants would first draft Philip Rivers—considered the second-best quarterback prospect in the draft—and then swap him for Manning *plus* give up their third-round pick (number 65) that year as well as their first- and fifth-round picks in the 2005 draft. Option 2 was to trade down with the Cleveland Browns, who held the seventh pick and also wanted a quarterback. At number seven, the Giants probably would draft the consensus third-best quarterback in the draft, Ben Roethlisberger. In exchange for moving down, the Giants would also receive from Cleveland their second-round pick (number 37) that year.

The Giants chose the first option, which meant they effectively considered Eli Manning to be worth more than Ben Roethlisberger plus *four additional players*. It turns out that this matched the chart perfectly.

To the two economists, however, this seemed like an extraordinarily steep price. They wondered whether perhaps the circumstances were exceptional; perhaps Manning's extraordinary pedigree reduced risk and made him a special case. But after collecting data from the NFL draft over the previous 13 years and looking at trades made on draft day as well as the compensation—salaries plus bonuses—paid to top picks, they found that the Manning trade was anything but unusual. As a matter of routine, if not rule, teams paid huge prices in terms of current and future picks to move up in the draft. They also paid dearly for contracts with those players.

In Manning's case, not only did he effectively cost the Giants four other players—one of whom turned out to be All-Pro linebacker Shawne Merriman—he was also given a six-year $54 million contract. Compare this to Roethlisberger, ultimately drafted eleventh by the Pittsburgh Steelers, who received $22.26 million over six years. Massey and Thaler found that historically, the number one pick in the draft typically is paid about 80 percent—80 percent!—more than the eleventh pick on the initial contract.

Massey and Thaler also found that the inflated values teams were assigning to high picks were remarkably, if not unbelievably, consistent. From used cars to commodities to real estate, markets inevitably vary. After all, different people with different needs and different resources make different valuations, and you'd think the market for football players, inherently subjective and speculative, would be *especially* erratic. But when it came to NFL draft picks, virtually every team agreed on the same values. No matter the circumstances or a team's needs, teams routinely assigned the same value to the same pick. Why?

It turned out they were all using the chart Mike McCoy created in 1991!

To test their suspicions that the chart overvalued high picks, Massey and Thaler compared the values teams placed on picks—either in terms of the picks and players they gave up or in terms of compensation—with the actual performance of the players. The economists then compared those numbers with the performance of the players given up to get those picks. For example, in the case of Eli Manning, how did his performance over the next five years compare with that of Philip Rivers *plus* the performances of the players chosen with the picks the Giants had to give San Diego to get Manning? Likewise, how did those numbers stack up against Ben Roethlisberger's stats *and* those of the players the Giants could have had with the additional picks they would have received from Cleveland?

More generally, if the chart says the number one pick will cost you the number six pick plus the number eight pick, if the chart is right, the performance of the number one player drafted should be the same as the total performance of the number six and number eight picks combined. The economists looked at the probability of making the roster, the number of starts, and the likelihood of making the Pro Bowl.

They found that higher picks are better than lower picks on average and that first rounders on average post better numbers than do second rounders, who in turn post better stats than third-round draft picks, and so on. No one will try to tell you that collectively first-round picks do not end up as better pros than third-round picks or that third-round picks don't outperform sixth-round picks.

The problem was that they weren't *that* much better. For example, according to the chart, the number one pick in the first round should be worth roughly five times the thirty-third pick, that is, the first pick in the second round. But it turns out that the top pick on average is not even twice as good as the thirty-third-picked player, yet teams pay the number one pick four to five times more than the thirty-third player drafted. Even within the first round, the chart claims that the number ten pick is worth less than half as much as the number one pick and accordingly is paid about half as much. But in reality, the typical number ten pick is almost as good a player as the typical number one pick.

Even looking position by position, the top draft picks are overvalued. How much better is the first quarterback or receiver taken than the second or third quarterback or receiver? Not much. The researchers concluded the following:

- *The probability that the first player drafted at a given position is better than the* second *player drafted at the same position is only 53 percent, that is, slightly better than a tie.*
- *The probability that the first player drafted at a position is better than the* third *player drafted at the same position is only 55 percent.*
- *The probability that the first player drafted at a position is better than the* fourth *player drafted at the same position is only 56 percent.*

In other words, selecting the consensus top player at a specific position versus the consensus fourth-best player at that position increases performance, measured by the number of starts, by only

6 percent. And even this is overstating the case, since the number one pick is afforded more chances/more starts simply because the team has invested so much money in him. Yet teams will end up paying, in terms of both players and dollars, as much as four or five times more to get that first player relative to the fourth player. If we look back at the 2004 NFL draft, was Eli Manning really 50 percent better than Philip Rivers and twice as good as Ben Roethlisberger? We could debate the ranking among those three today. Putting aside Roethlisberger's troubling and well-chronicled "character issues," most experts and fans probably would rank them in reverse order from their draft selection in terms of value today. You'd be hard put to convince anyone that Manning is appreciably more valuable than Rivers or Roethlisberger; in any event, he's certainly not *twice* as valuable. Yet this pattern persists year after year.

Is having the top pick in the NFL draft such a stroke of good fortune? It's essentially a coin flip, but not in the traditional sense. Heads, you win a dime; tails, you lose a quarter. Massey and Thaler go so far as to contend that once you factor in salary, the first pick in the entire draft is worth *less* than the first pick in the second round. (For kicks, imagine the team with the top pick showing up on draft day, the fan base brimming with exuberant optimism, only to hear the commissioner intone: "With the first pick, the Detroit Lions . . . pass. The Cleveland Browns are now on the clock.")

Massey and Thaler discovered another form of overvaluation as well: Teams paid huge prices in terms of *future* draft picks to move up in the draft. For example, getting a first-round draft pick this year would cost teams two first-round picks the next year or in subsequent years. Gaining an additional second-round pick this year meant giving up the first- and second-round picks next year. Coaches and GMs seemed to put far less value on the future. Looking at all such trades, the so-called implicit discount rate for the future was 174 percent, meaning that teams valued picking today at more than twice and nearly three times the value of taking the same pick in the next year's draft! Think of this as an interest rate. How many of us would borrow money at an annual interest rate of 174 percent? Even loan sharks aren't that ruthless.

■ ■ ■

Why do NFL teams place so much value on high picks? Psychology explains a lot of it. As anyone who's ever watched the game show *Deal or No Deal* or has placed a bid on eBay or at a charity auction can attest, we tend to overpay for an object or a service when we're in competition with other bidders. We know the value of that $500 gift certificate and raise our paddles accordingly. That's easy. What if the value of the item is uncertain? When a coveted piece of art or jewelry is up for grabs and the value is unclear, the real bidding war begins, often resulting in overpayment. There's even a term for this: winner's curse.

To demonstrate the winner's curse, a certain economics professor has been known to stuff a wad of cash into an envelope, stating to his students that there is less than $100 inside. The students bid for its contents; without fail, the winning bid far exceeds the actual contents—sometimes even exceeding $100! (The surplus is used to buy the rest of the students pizza on the last day of class.) The top picks in the NFL draft, by its very nature an exercise in speculation, are singularly ripe for the winner's curse.

Here's another factor in the overpayment of draft picks: As a rule, people are overconfident in their abilities. In all sorts of different contexts, we're more sure of ourselves than we probably should be. In a well-known study, people were asked at random whether they were above-average drivers. Three-quarters said they were. Similarly, between 75 and 95 percent of money managers, entrepreneurs, and teachers also thought they were "above average" at their job. Not everyone, of course, can be above average. You'd expect roughly half to be. How many times are hiring decisions made on intuition because a boss on the other side of the desk is convinced that in a 30-minute interview she's found the best candidate? How often do doctors advise a treatment plan because of intuitive decision-making, not because of evidence-based decision-making? They're just sure they're right. In the same way,

NFL general managers (and sometimes their interfering owners) tend to be overconfident in their ability to assess talent. They trust their gut. Not altogether a bad thing, but they trust it too much and overpay as a result. Never mind the math; they fancy themselves the exception. They *know* they're right about this player, just as every entrepreneur knows his business plan is better and every mutual fund manager *knows* she's got the winning stock picks.

Overvaluing because of gut instincts also helps explain why teams invest so much in this year's pick and so little in next year's prospects. The guy in front of you is "a once-in-a-lifetime player," a term invoked almost without fail at every draft. The guy next year is just an abstraction. (Another explanation for the immediacy: GMs and coaches typically have short tenures, so winning now is imperative to keeping their jobs. Even a year can seem beyond their horizon.)

Overvaluing draft picks isn't confined to football. In the NBA, teams value this year's pick at two to three times the value of the same pick in next year's draft. Collecting data on the NBA draft going back to 1982 and looking at trades for current draft picks that involved future draft picks, we found that future draft picks were discounted heavily at 169 percent, almost to the same extent that NFL teams discounted future draft choices. Again, this could be because GMs and coaches have short windows or because teams overvalue what they see now and undervalue what they can't see readily. Top draft picks in the NBA, however, were only slightly overvalued—not nearly to the extent they are in the NFL. This makes sense: The NBA, after all, has only two draft rounds. Player ability is also easier to predict, there are fewer players and fewer positions to consider, and a single player has a much larger impact on the team than in football. In baseball—where the draft is less important, as so many foreign players sign as free agents—there is also a huge discount applied to future picks, and top picks again tend to be overvalued.

The truth is that evaluating talent is *hard*. How hard? For an illustration, consider the case of Eli Manning's older brother, Peyton. In 1998, Peyton Manning entered the NFL draft with tremendous hype. But teams weren't sure whether he would be the first or second quarterback taken. There was a comparably touted quarterback from Washington State, Ryan Leaf, considered by many NFL scouts to be the better prospect. Leaf was bigger and stronger than Manning, two easily measurable characteristics, and, again with the support of numbers, was regarded as the better athlete. Although Manning acquitted himself capably at Tennessee, he never led the formidable Volunteers to a national title, a cause for some concern.

The San Diego Chargers originally held the third pick of the draft but made a trade with the Arizona Cardinals to move up to the second pick to ensure that they got one of the two tantalizing quarterbacks. This move cost them two first-round picks, a second-round pick, reserve linebacker Patrick Sapp, and three-time Pro Bowler Eric Metcalf—all to move up one spot!

In the end the Indianapolis Colts, holding the number one pick, took Manning. The San Diego Chargers took Leaf with the second pick and signed him to a five-year contract worth $31.25 million, including a guaranteed $11.25 million signing bonus, at the time the largest ever paid to a rookie—that is, until the Colts paid Manning even more: $48 million over six years, including an $11.6 million signing bonus.

You probably know how the story unfolded. Peyton Manning is the Zeus of NFL quarterbacks, a four-time MVP winner (the most of any player in history), a Super Bowl champion, riding shotgun on the express bus to the Hall of Fame. And Leaf? As we write this, he is currently out on bond as he defends himself against burglary and drug charges in Texas. (He was sentenced to probation after pleading guilty to illegally obtaining prescription drugs.) In the summer of 2009, he was arrested by customs agents as he returned from Canada, where he had been in drug rehab. He played his last NFL game in 2001, his career marked by ineffective play; injuries; toxic relations with teammates, coaches, and the media; and a general lack of professionalism. Leaf once complained of wrist pain to avoid practice but reportedly had played golf earlier in the day. Another time, while serving a four-game suspension for insubordination and ordered to rehabilitate a shoulder and wrist, he was videotaped playing flag football with friends.

After three disastrous seasons, the Chargers released Leaf, which actually broke with convention. Time and again, ignoring what economists call sunk costs, when a high draft pick underperforms, teams tend to keep investing in him for years in the hope that he'll turn it around. The team rationalizes: We paid a ton to get him; we have to keep trying to make it pay off. But the money's gone, and continuing to invest in the underperforming player is simply making a bad decision worse. Ever ordered a food item, bitten into it, and found it tasted awful? How many of us eat it anyway because, well, we paid for it? Or how often do we insist on holding on to a stock we bought for $50 that is worth $40 today and $30 tomorrow? Is the off-tasting sushi really going to taste better if we keep eating it? Is the company that is tanking really going to see its stock price return to $50? The money is gone; we may as well cut our losses now. But we hardly ever do. Everyone hates admitting a loss, football executives included.

To the credit of the Chargers organization, they learned from their costly mistake. In 2000, still recovering from the Leaf fiasco, the Chargers finished 1–15 and were "rewarded" with the top pick in the 2001 NFL draft. They traded the pick to Atlanta for the Falcons' number five as well as a third-round pick, a second-round pick in 2002, and Tim Dwight, a wide receiver and kick return specialist. The Falcons used the top pick to select Michael Vick; the Chargers used that fifth selection on LaDainian Tomlinson, who would become the most decorated running back of his generation. To satisfy their quarterback needs, they waited until the first pick of the second round and tapped Drew Brees, who would go on to become the 2010 Super Bowl MVP, albeit for a different team, the New Orleans Saints. Remember, too, the Chargers were willing to give up Eli Manning in 2004 for what was effectively Shawne Merriman and Philip Rivers. Add up the values from both trades and they follow McCoy's chart almost exactly. Imagine how much better the Chargers could have done if they'd known the chart was flawed.

Beyond the money, overinvestment in high draft picks can have other real costs. Pampering the first-round pick—treating him differently from the sixth-rounders who'd be put on waivers for a comparably dismal performance—exacts a price on team performance and morale. It also forestalls taking a chance on another athlete. The more chances given Ryan Leaf, the fewer chances afforded his backup. And it's not just the team that has drafted the player that's prone to this fallacy. Even after Leaf's miserable performance and behavior in San Diego, three other teams gave him another shot. They recalled the player he was in college. They still coveted his size, strength, and athleticism and believed the hype. Never mind the clear evidence that he was a bust. "It'll be different here," they told themselves. Only, of course, it wasn't.

■ ■ ■

How do teams know when they're getting the next Peyton Manning and when they're getting the next Ryan Leaf? They don't. There are simply too many unknowns and too much uncertainty to know whether you've drafted a great player or a bust. The only certainty is that you will pay dearly for both. Manning and Leaf were both very expensive, but only one of them was able to perform. Also, the uncertainty regarding whether Manning was better than Leaf is not uncommon. Take a look at the following table and compare the number one picks over the last decade with the players voted offensive and defensive rookies of the year in the subsequent year as well as other players at the same position available in that draft who made the Pro Bowl.

Still think it's easy to pick the best players? If you look at the top picks in the NFL draft from 1999 to 2009, not a single one was named rookie of the year on either side of the ball. More damning, many of the top picks have turned out to be busts. Of the last 11 number-one-picked players, eight have been quarterbacks. Four of them—Tim Couch, David Carr, Alex Smith, and the beleaguered JaMarcus Russell—came nowhere close to justifying the selection. Of the four remaining quarterbacks, it's too early to tell what will become of Matthew Stafford in Detroit,

TOP DRAFT PICKS FROM 1999 TO 2009
ROOKIE OF THE YEAR

Year	#1 Pick	Offensive	Defensive	Pro Bowl players from same draft at same position
2009	Matthew Stafford (QB)	**Percy Harvin (#22)**	**Brian Cushing (#15)**	N/A
2008	**Jake Long (OT)**	Matt Ryan (#3)	Jerod Mayo (#10)	Ryan Clady (2nd OT)
2007	JaMarcus Russell (QB)	**Adrian Peterson (#7)**	**Patrick Willis (#11)**	N/A
2006	**Mario Williams (DE)**	**Vince Young (#3)**	**DeMeco Ryans (#33)**	Elvis Dumervil (9th DE)
2005	Alex Smith (QB)	Cadillac Williams (#5)	**Shawne Merriman (#12)**	Aaron Rodgers (2nd QB), Derek Anderson (11th QB)
2004	**Eli Manning (QB)**	**Ben Roethlisberger (#11)**	**Jonathan Vilma (#12)**	Phillip Rivers (2nd QB), Ben Roethlisberger (3rd QB)
2003	**Carson Palmer (QB)**	**Anquan Boldin (#54)**	**Terrell Suggs (#10)**	Tony Romo (14th QB, undrafted)
2002	David Carr (QB)	**Clinton Portis (#51)**	**Julius Peppers (#2)**	David Garrard (4th QB)
2001	**Michael Vick (QB)**	Anthony Thomas (#38)	**Kendrell Bell (#39)**	Drew Brees (2nd QB)
2000	Courtney Brown (DE)	Mike Anderson (#189)	**Brian Urlacher (#9)**	Shaun Ellis (2nd DE), John Abraham (3rd DE), Kabeer Gbaja-Biamila (12th DE), Adewale Ogunleye (24th DE, undrafted)
1999	Tim Couch (QB)	**Edgerrin James (#4)**	**Jevon Kearse (#16)**	Donavan McNabb (2nd QB), Daunte Culpepper (4th QB)

Bold indicates made the Pro Bowl at any point in career.

and though Carson Palmer and Michael Vick have each been to the Pro Bowl, both have also spent considerable time on the sidelines, Palmer because of a gruesome knee injury and Vick because he was incarcerated for nearly two years and suspended from the league for his involvement in an illegal dog-fighting scheme. That leaves only one number one quarterback pick, Eli Manning, who has started the majority of games for his team since his debut.

But again, divergent as their careers have been, all the number one picks were paid handsomely. So for the teams selecting at number one, the best-case scenario is that you get a good player for an expensive price. You buy a Camry at Porsche prices. Worst-case scenario, you pay a lot of money and get nothing in return. You pay the price of a Porsche for a clunker. What you will never get is a great player at a cheap price. You never get the Porsche at the clunker price in the early rounds.

In the 2010 draft, the trend continued as the St. Louis Rams selected quarterback Sam Bradford with the number one overall pick (making 9 of the last 12 first picks QBs) and promptly signed him to the richest contract in history—five years at $86 million with $50 million guaranteed.

Even with successful high picks on the order of Manning and Palmer, the question isn't how much they cost in terms of salary but also how much they cost in terms of the draft picks you could have taken instead. In 2005, the San Francisco 49ers drafted quarterback Alex Smith with the first pick; the only quarterbacks from the 2005 draft to have made a Pro Bowl are Aaron Rodgers (number 24 pick), who made it for the first time in 2009, and Derek Anderson, the eleventh quarterback taken that year. In 2000, defensive end Courtney Brown was chosen as the number one pick. He never made a Pro Bowl. But Shaun Ellis and John Abraham, the second and third defensive ends taken in that draft, did make numerous Pro Bowls. As did Kabeer Gbaja-Biamila, the twelfth defensive end taken that year with the 149th pick, and Adewale Ogunleye, who wasn't even drafted that year, meaning at least 24 defensive ends were chosen before him. Bottom line: In football, it's very hard to tell who is going to be great, mediocre, or awful.

So what should a team do if it's blessed (which is to say, cursed) with a top pick? Trade it, as the San Diego Chargers learned to do. Drafting number 10 and number 11 instead of number 1 is a much better proposition. The Dallas chart shows that the value of these players should be the same, but the reality is quite different. Teams get far more value from having picks at number 10 and number 11 than they do by taking a chance on one pick at number one. With two "draws," the chance of having at least one of the two picks succeed is much higher, and the cost is the same or less. Factor in the potential for injuries and off-the-field trouble and it becomes even more apparent that having two chances to find a future starter is a much better proposition than having only one. Also, the team avoids the potential of a colossal and public bust like Ryan Leaf. Even if the later picks flop, fans won't care nearly so much as they will when the top pick is a bust.

■ ■ ■

Perhaps NFL owners have different objectives, but it's safe to assume they want to win or make money, and probably both. Following the chart meant that they lost on both counts. They overvalued top talent and, even when a pick happened to pan out, paid dearly.

For a franchise willing to ignore convention and depart from the chart (or improve it), the payoff can be huge. A team that discovered the chart was flawed—that it overvalued top draft picks—could trade its high picks for many more lower picks. It wouldn't be taking the sexy picks and exciting fans by drafting Heisman Trophy winners and standouts at the NFL combine, but as Massey and Thaler's research shows, it would field better teams and win more games. If you look at the teams that did trade down in the draft or traded current picks for a greater number of future picks, the researchers showed that those teams improved their winning percentages significantly over the four years after each trade.

Over the last decade, two teams in particular went "off the chart," as it were, and created a new model, placing less value on the top picks: the New England Patriots and the Philadelphia Eagles. Not surprisingly, those two teams have two of the top winning percentages and five Super Bowl appearances between them since 2000. Tom Brady, one of the few quarterbacks hailed as Peyton Manning's equal, a former MVP and three-time Super Bowl winner? He was drafted in the sixth round of the 2000 draft with the 199th pick and thus was obtained cheaply, providing the Patriots extra cash to collect and keep other talent to surround him. Teams that traded current draft picks for future ones benefited in subsequent years, too, and again, the Patriots and Eagles were at the forefront. (Not coincidentally, their coaches, Bill Belichick and Andy Reid, have enough job security to afford the luxury of a long-term focus. Reid even has the additional title of executive vice president of football operations.)*

Which teams are on the other end of the spectrum, routinely trading up in the draft to get higher picks and overpaying for them? The answer is unlikely to surprise you: the Oakland Raiders and the Washington Redskins, who collectively have the fewest number of wins per dollar spent.†

As for the Cowboys, as we write this, their quarterback is All-Pro Tony Romo, who currently owns one of the highest passing ratings of all time. Never mind the Cowboys fleecing other teams with a pricing system for draft picks. After starring at tiny Eastern Illinois University, Romo wasn't selected at all in the draft, so Dallas simply acquired him as a rookie free agent in 2003 and nourished him. One of Romo's favorite targets, wide receiver Miles Austin, was also undrafted when he left tiny Monmouth University. He, too, was spotted by the Cowboys and signed as a rookie free agent. It seems the Cowboys may have found other ways to find value among new players outside the draft, deviating from the system they created.

---

* Bill Belichick and Andy Reid are also two of the least conventional coaches in terms of their play-calling (both go for it on fourth down more often than average) and are, of course, routinely criticized for it, especially when it fails.
† Ironically, the Redskins had extensive discussions with Massey and Thaler early on in their research, and the two professors met with team owner Dan Snyder and his football staff. After receiving the advice to go off the chart and trade down in the draft and give up current picks for future ones, the Redskins did exactly the opposite.

HOW A COIN TOSS TRUMPS ALL

Why *American Idol* is a fairer contest than an NFL overtime

It's one of the great ironies in sports. For 60 minutes, the gladiators in the NFL risk life and head trauma, bouncing off one another, driving opponents into the ground, and generally purveying violence and mayhem. They're caked in blood and sweat and dirt and grass. It's all part of the spectacle that makes professional football so indefensible to some and so compulsively watchable to the rest of us.

And then, if the two teams are tied after regulation, these fierce and brutal struggles are decided largely by . . . the flip of a coin. After that gamelong physical exertion, the outcome ultimately comes down to dumb luck. Sure, the winning team has to kick the ball through the uprights or, in rare cases, march into the end zone, but that's mostly a formality. Win that arbitrary coin toss at midfield and earn first possession of the ball in overtime, and victory is usually yours. As the broadcaster Joe Buck once quipped before the Oakland Raiders and San Francisco 49ers were about to begin overtime: "[Here comes] one of the biggest plays of the day, the coin flip!"

Credit Brian Burke, an aerospace engineer, a former F/A-18 carrier pilot in the U.S. Navy, and the current overlord of the website advancednflstats.com, for logging the hard miles here. Burke determined that between 2000 and 2009, 158 NFL games, including the playoffs, went to overtime. Two of those games ended in a tie. In one game, the Detroit Lions won the coin flip and chose neither to kick nor to receive but rather what side of the field they preferred to defend. (Not surprisingly, they lost.) In the other 155 games, the team that won the coin flip won the game 96 times, a 61 percent clip. As Burke correctly pointed out on his website, "Don't be tricked by people that say 'only 61 percent.' If we agree 50 percent would be the fairest rate, you might think 61 isn't very far from 50. But that's not the right way to look at it. The appropriate comparison is 61 percent versus 39 percent, the respective winning percentages of the coin flip winners and losers. That's a big advantage—over 3:2 odds."

What's more, in 58 of the 158 games, or 37 percent of the time, the team that won the coin flip won the game in its first possession. Think about this for a second: Teams battled their guts out for 60 minutes over four quarters and were tied with their opponents. Then, 37 percent of the time, one team lost in overtime *without even touching* the ball. Is it any wonder that of the 460 coin toss winners in NFL history, only 7 of them have elected to kick off and play defense first in overtime play?

David Romer, the Berkeley economist who encouraged more teams to go for it on fourth down, has a way to make NFL overtimes fairer: Change the spot for the initial kickoff. As it stands now, the kicking team boots from the 30-yard line. At this distance, it's difficult to kick the ball into the end zone for a touchback, so the receiving team often gets the chance for a strong return. Romer claims that moving the kickoff up just five yards to the 35 would trigger a significant increase in touchbacks so that the receiving team would begin at the 20-yard line, the "break-even point" where the team on offense and the team on defense are equally likely to score next.

Chris Quanbeck, an electrical engineer and rabid Packers fan, offers a more radical, and intriguing, suggestion: Auction off the first possession of overtime, using field position as currency. Want the ball first? How far back are you willing to start your first drive? If we accept Romer's premise that the 20-yard line is the break-even point, if Team A is willing to start on its own 15-yard line, Team B might happily agree to start out on defense. Writing in *Slate,* Tim Harford, a columnist for the *Financial Times,* noted this additional benefit: "Imagine the possibilities for stagecraft. . . . The two head coaches could come to midfield with sealed bids, with the envelopes to be opened by a cheerleader representing each team—a gridiron version of *Deal or No Deal.*"

Others have suggested eliminating field goals in overtime and mandating that the winning team must score a touchdown. As kickers have improved their accuracy and leg strength, it's become increasingly easy for the team that wins the coin toss to reach field goal range. What's more, eliminating the field goal would encourage the kicking team to inaugurate the overtime by attempting an onside kick. (As it stands, it's a foolish play. If the receiving team recovers, they're likely to be only 10 or 15 yards away from being in position for a game-winning field goal.)

The other obvious solution would be to adopt some variation of the "Kansas Plan" in college football, whereby each team receives a first-and-ten possession at the opponent's 25-yard line. Here, at least the team unlucky on the coin toss gets the equivalent of "last licks"—it can't lose without touching the football on offense.

In response to an Internet discussion, one reader suggested simply letting the fans vote for the winner, in the manner of *American Idol.* The poster was being facetious, of course. But is it that much more ridiculous than deciding a game—a tightly contested game at that—largely on the basis of heads or tails?

There are other sports that employ the flip of a coin, if not to such dramatic effect. In professional tennis, the winners of the prematch coin flip have four choices: They can serve, return, choose one side of the court, or forgo the choice entirely. The overwhelming majority of players opt to serve first, and this makes sense. In ATP matches in 2009, servers won 78.4 percent of the time.

At the 2010 NFL owners' meetings, the league passed a change to the overtime rule in playoff situations. The team losing the coin toss will have a chance to score if the opposing team kicks a field goal. But if the team that wins the coin toss scores a touchdown on its first possession, the game will be over. If both teams exchange field goals, sudden death commences, with the first team to score again (even if it's just a field goal) winning. Teams voted the modification in by a margin of 28–4, in part because of the data showing how often the initial random overtime coin flip determines the game's outcome. "Plenty of people on the committee, myself included, are so-called traditionalists," Bill Polian, the Indianapolis Colts' president, told reporters. "I am proud to be one. But once you saw the statistics, it became obvious we had to do something."

At this writing, the change will be implemented only for playoff games, when a winner *must* be determined. During the regular season, teams that are tied at the end of regulation time will continue to toss a coin to see who receives the ball in overtime, and the team that is luckier probably will walk off the field victorious.

WHAT *ISN'T* IN
THE MITCHELL REPORT?

Why Dominican baseball players are more likely
to use steroids—and American players
are more likely to smoke weed

On January 11, 2010, Mark McGwire, the former St. Louis Cardinals slugger, attended confession. He sat across from Bob Costas on an MLB Network set made to look like a cozy living room—replete with lamps, urns, and a faux fireplace—to talk, finally, about the past. He looked noticeably less bulked up than the behemoth who'd captivated the country during his 1998 pursuit of Roger Maris's single-season home run record. His familiar red hair now salted with gray, McGwire stifled tears and looked shamefaced as he confirmed what most baseball fans had long suspected. Yes, he admitted, he had used performance-enhancing drugs, or PEDs. He then asked for forgiveness.

As apologies go, McGwire's did not exactly set a new benchmark for sackcloth-and-ashes contrition. His mea culpa had been orchestrated by Ari Fleischer, George W. Bush's first White House press secretary, now a sports consultant and crisis manager. In the weeks leading up to the interview, Fleischer prepped McGwire on every conceivable question he might face. "It was just like batting practice," explained McGwire. "[The] attitude was: You're not going to get blindsided."

McGwire's confession coincided with an offer to become the St. Louis Cardinals' hitting coach. (Was his admission a crisis of conscience or a condition for a new job?) Interspersed among the pleas of penance, McGwire lamented, "I wish I had never played in the steroid era," as if he'd had no say in the decision to juice up and had simply, by accident of birth, had the misfortune of playing at the wrong time. He also echoed the increasingly familiar explanation of countless other athletes caught in the steroids web: "The only reason that I took steroids was for my health purposes. I did not take steroids to get any gain for any strength purposes."

Although McGwire later reflected that his confession "went wonderfully," public opinion was split. In the minds of most fans, there is enough circumstantial evidence to convict Roger Clemens, Barry Bonds, and Sammy Sosa of steroid use, too. McGwire was the first from that group to come forward voluntarily and make an outright admission of guilt. Good for him. But McGwire's insistence that he'd used the drugs only for recovery from an injury rang hollow. Steroids are a performance-enhancing drug, and it is no coincidence that McGwire's biggest years in terms of home run production coincided with the period in which he now admits he juiced up. If McGwire had used the drugs only for convalescence and not for strength, why had he felt compelled to apologize to the Maris family? McGwire's critics, unmoved by his apology, contended that he'd become the face of the steroids era in baseball.

■ ■ ■

For most baseball fans, steroids are commonly associated with Major League stars like McGwire, José Canseco, Roger Clemens, Alex Rodriguez, and Manny Ramirez: sluggers with chests and arms as disproportionately large as their home run totals, ageless pitchers throwing heat into their forties, and utility infielders suddenly jacking 30 home runs in a season. Those were the cheaters who distorted the competition and tainted baseball's hallowed statistics, who made it into former U.S. senator George Mitchell's report.* Those were the players who made steroids such a cause célèbre.

But that picture is inaccurate. Most of the steroids in baseball were purchased and consumed by players whose names won't be familiar to even the most die-hard fans and who are not listed in the Mitchell Report. Among the 274 professional players who tested positive for steroids and other banned performance-enhancing drugs between 2005 and the fall of 2010, 249, the overwhelming majority, were *minor league* players whose faces never made it onto the front of a baseball card.

The true face of the steroid era? It might look a lot like the smiling mug of Welington Dotel. An irrepressible outfielder with a lively bat and an arm so powerful that it should require a license, Dotel grew up in Neiba, Dominican Republic, a fairly nondescript town in the southwestern quadrant of the island, not far from the border of Haiti. Welington was the oldest of five children born to parents who kept the family afloat by doing a series of odd jobs: working in restaurants, working on roads, teaching. According to Welington, "It changed with the season, but they did many things."

The family struggled in a region where average annual household income was less than $9,000. "We were not rich," Welington says, laughing. Yet he was better off than some of his friends and neighbors, who played baseball with milk cartons for gloves and sticks for bats. As with so many boys on the island, he fell hard for baseball. The game fed something inside him. But it was also a way, he dreamed, to deliver his family from poverty.

Genial and outgoing, Welington makes friends easily. His for-

---

* The "Report to the Commissioner of Baseball of an Independent Investigation into the Illegal Use of Steroids and Other Performance Enhancing Substances by Players in Major League Baseball" was the culmination of a 21-month investigation that produced the 409-page report released on December 13, 2007. The report named 89 players alleged to have used PEDs.

mal education ended before high school, when, like most Dominican prospects, he dropped out to pursue a career on the ball field.
Asked what he'd be doing were it not for baseball, he pauses.
"Maybe teaching baseball," he says. "Something with baseball
because it's my passion. Maybe even more than my passion. It's
everything to me."

A late bloomer, Welington was 18 when he was signed as a free
agent by the Seattle Mariners organization. His signing bonus, he
says, was $160,000. "It was unbelievable," he says. "They told
me and I was like, 'Sure, I sign that!'" He bought his mom a new
home, financed a new car, and acquired some of the other material possessions no one in his family had ever owned.

When Dotel reported to the Mariners, it marked the first time
he'd left the Caribbean. He staved off homesickness by remem-
bering just how lucky he was to have the opportunity. His professional career started auspiciously enough. He returned to his
island in 2005 and played for the Mariners team in the Dominican
Summer League, hitting .373 in 69 games. But in his first year of
Rookie ball, his career stalled a bit. Playing in Peoria, Arizona,
2,500 miles (and immeasurable cultural miles) from home, Dotel
hit .261 with seven home runs. Not terrible at age 20 but not the
kind of numbers that impress the franchise gatekeepers.

Then, toward the end of the 2006 season, he tested positive for
an undisclosed banned performance-enhancing substance. He was
issued a 50-game suspension, which he served the next spring. It
is, understandably, not his favorite topic of conversation. He'd
rather not discuss, for instance, whether he thought the testing
procedure was fair or how long he had used the banned substance.
"We make mistakes when we're young, and we try to learn from
it," he says.

When asked who "we" are, he replied, "Young players. Young
players like me."

Ozzie Guillen, the famously candid Chicago White Sox manager and a native of Venezuela, echoed this sentiment in 2010
when he told reporters: "It's somebody behind the scene making
money off those kids and telling them to take something they're

not supposed to. If you tell me, you take this and you're going to be Vladimir Guerrero or you're going to be Miguel Cabrera, I'll do it. Why? Because I have seven younger brothers that sleep in one bed in the same room. I have to take care of my mother, my dad." Then Guillen added: "No, no dad. Two guys got dad." (In other words, another common trait among these Latin players is that they have absentee fathers or are in single-parent families.)

Though the gears of globalization are unmistakably rotating, America's pastime is still populated mostly by Americans. Of the nearly 1,600 players on Major League rosters between 2005 and 2010, nearly three out of four were from the United States. But when we constructed a database for baseball players who had been suspended for PEDs, we noticed there was only one player with the surname Smith and *seven* with the surname Rodriguez.

It turns out that although American players account for 73.6 percent of those in the Major Leagues, they represent just 40 percent of suspended drug offenders—about half as many as one would expect if drug suspensions were simply proportional to demographic representation. In contrast, Dominicans represent barely 10 percent of Major League players but account for 28 percent of the drug suspensions—more than two and a half times more than their representation would indicate. Venezuelans account for about 6 percent of all Major League players but more than 12 percent of those suspended for drugs. In other words, the numbers suggest that a Dominican or Venezuelan player is about *four times* more likely to face suspension for using PEDs than his U.S. counterpart.

The disparity is even more pronounced in the minor leagues. Among all 8,569 minor league baseball players in the United States from 2005 to 2009,* American-born players are less than half as likely to test positive as their numbers would have indicated. And once again, Dominican and Venezuelan players are two to three times more likely to appear on the drug suspension

* We look only at players in the U.S. minor leagues (Rookie, A, Double-A, and Triple-A leagues) and exclude any suspensions from amateur leagues in other countries, as those leagues may have different drug and enforcement policies.

list than their proportion of the total population of minor league players. Overall, players from the Dominican Republic and Venezuela are more than four times more likely to test positive for banned performance-enhancing drugs than their U.S. counterparts in the minor leagues. Players born in Colombia, Cuba, Mexico, and Puerto Rico are overrepresented, and players from Australia, Canada, and Japan and Taiwan are similarly underrepresented.

The graph below summarizes the data on PED use in baseball by country of origin. The vertical line shows the baseline, with positive tests being perfectly proportional to players from that country. Countries whose numbers go beyond the line are overrepresented, and those with numbers short of the line are underrepresented.

PERCENTAGE OF PED SUSPENSIONS RELATIVE TO PERCENTAGE OF PLAYERS IN MAJOR AND MINOR LEAGUES BY COUNTRY OF ORIGIN

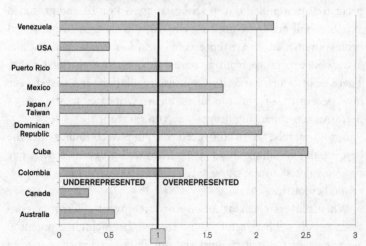

Likelihood of Being Caught Using PEDs Relative to Population

Data for 1,520 U.S. Major League and 8,569 U.S. minor league players from 2005 to 2009.

Why this huge disparity? There are a host of plausible explanations for this complex and thorny issue, touching on everything from culture to education, and we don't discount any of them.

Perhaps American players are just as likely to cheat but lack access to steroids, which are legal in some countries. Maybe American players use steroids just as frequently but are less likely to get caught because they are more adept at masking their use or cycling off their regimens before testing. American players might be better educated about the health and psychological risks, the baseball drug regulations, and the nuances of the testing protocol. More cynically, perhaps Dominicans, Cubans, and Venezuelans are targeted more by the drug testers and are therefore more likely to be caught.

One could also argue that comparing Latin players to American players is an inherently flawed exercise, since the ways they come to baseball bear little resemblance. At the moment (though this is likely to change in the near future, in large part to address the corruption), Dominican and Venezuelan players aren't subject to the Major League draft. Instead, they are independent free agents who are signed by teams through recruitment. Bright prospects are often pulled out of school at age 13 or 14 and groomed at a baseball academy until they're 16, old enough to be signed professionally. Most are represented by a *buscon,* a local middleman who can range from a trusted adviser to a glorified pimp. There is little infrastructure and less regulation. It's easy to see how deception—from fraudulent birth certificates to the use of performance-enhancing drugs—can run rampant.

But ultimately, we're convinced that simple economics does the best job of explaining why players from well-off countries are five times less likely to test positive for PEDs than players from impoverished countries.

When players cheat or deceive or circumvent rules, they consider a trade-off between risk and reward, balancing the potential advantage of the gain against the possibility and cost of getting caught and the punishment they would face, whether money, guilt, or condemnation by peers. This balancing is true of all of us. Every decision involves considering two kinds of cost—the cost of taking an action versus the cost of not taking action—an "opportunity cost." We might be disinclined to blow through a stop sign in the middle of the day when other cars are at the intersection and a traffic cop might be stationed nearby. In the middle of the night with no one else in sight, we might arrive at a different decision. We might park illegally if the penalty is a $50 ticket. We're less likely to do so when the ticket will cost us $500. Simple risk-reward analysis doesn't explain all of our actions—after all, rich people steal, too—but it explains a lot.

Under baseball's old testing program, the incentives to cheat were high for virtually all players, but they were particularly extreme for Dominicans and Venezuelans. Never mind the lure of a guaranteed Major League contract, which could ensure generational wealth. Consider Welington Dotel and his $160,000 Major League signing bonus. That is as much as his parents might make in decades of working. For many Dominican players, just making it to a national baseball academy that provided three square meals a day and a decent place to sleep represented a vast upgrade in their standard of living.

The risk of getting caught and failing a drug test was low. The punishment was low, as well. Under the old testing protocol, initial suspensions were only 15 games. As baseball cracked down on PED use after 2004, bowing to popular (and congressional) pressure, in a worst-case scenario, a positive test would trigger a 50-game suspension, still less than two months of the season. There was a huge upside and not too grave a downside. In economic terms, to dust off a tired sports cliché, players had little to lose and everything to gain. "You hate to even admit this," says one former National League scout. "But when you see how some of these kids grew up, part of you thinks they'd be nuts *not* to do everything they possibly could to make it, even if that means steroids."

And many do just that. To see the relationship between economic incentives and PED use among athletes, we plotted the likelihood of a player failing a drug test relative to the per capita gross domestic product (GDP) of his country of origin. The graph below shows that the probability of steroid use lines up almost perfectly with the wealth of the country. Players from Canada, Australia, and Japan are underrepresented as PED users, whereas

LIKELIHOOD OF USING PED IN U.S. MINOR LEAGUES BY COUNTRY OF ORIGIN

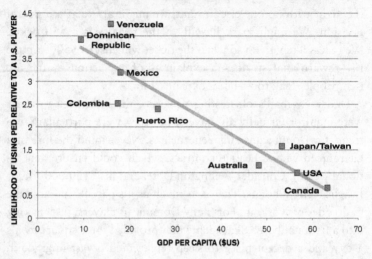

players born in Colombia, Cuba, Mexico, and Puerto Rico are overrepresented.

Of course, the wealth of countries as measured by GDP per capita not only captures the economic opportunities for its citizens but also is an indicator of a country's infrastructure, education, health, and many other factors that may influence an athlete's decision to improve his quality of life by boosting his career with performance-enhancing drugs. But virtually the same pattern occurs when we look at other measures of "opportunity," such as health, infant mortality, life expectancy, unemployment rates, literacy rates, schooling, access to clean water, percentage of population below the poverty line, and percentage of the population making less than $2 per day. In short, players from countries with lower standards of living and more limited opportunities are much more likely to use performance-enhancing drugs.

If economics are motivating drug use, a player from a poor country should be more likely to cheat from the very onset of his career. From day one, he is desperate to make it in baseball in light of the scarcity of other opportunities and the immediate financial impact of even that first signing bonus (all the more so if he owes a debt to his *buscon*). A player from a wealthier country probably wouldn't have the same economic incentive to use PEDs as a teenager: *Hey, if I'm not drafted so high, I have other options, like a college scholarship.* With athletes from wealthier countries, the economic incentive to do drugs would kick in later in a player's career. A Triple-A player facing his last chance to make it to the Majors, where the per diem alone would rival his current salary—*that* guy has an incentive to cheat. The thirtysomething Major League veteran looking for one last big-money contract? He, too, has an incentive to cheat. (Just note how many players cited in the Mitchell Report were veterans in the autumn of their careers.)

Sure enough, when we look by country of origin at the average age of players at the time they are caught using performance-enhancing drugs, we find a pattern that is consistent with economic incentives. The average age of players from the Dominican Republic and Venezuela caught using PEDs is 20 to 21. The average age of players from the United States who are caught is 27. (For players from Japan, Taiwan, Australia, and Canada, the average age is close to 30; for those from Mexico and Puerto Rico, it's 25.) The ages of those caught line up exactly with the wealth of their countries.

Could players from Venezuela and the Dominican Republic simply be younger on average than the U.S. players? Controlling for the average age of players from each country, we still find that steroid use among Venezuelan and Dominican athletes occurs much earlier than it does with their U.S. counterparts.

Here's still another indication of how economics underpin steroid use: Athletes more desperate to succeed will not only try to enhance performance from the very beginning of their careers but continue to use PEDs even if they've been caught. When we look at repeat offenders, we find that Venezuelans and Dominicans are vastly overrepresented.

Of course, the United States and other countries differ in many more ways than economic opportunities. Could cultural, educational, and moral issues play a role here? How do we know it isn't other differences, rather than economics, driving these findings? One way to answer this question is to look only at U.S.-born ballplayers. When we examined the hometowns and neighborhoods of American players who had failed a drug test,* we found that PED use is more prevalent among players who came from areas with lower average income, lower high school and college graduation rates, and higher unemployment—the same patterns we found across countries. Differences in culture, institutions, and governments are muted when we look only within the United States, but the economic motive remains.

What about ethics? Could it be that U.S. players are more upstanding? If that were the case, we might expect American players to obey drug rules across the board. Yet they don't. Looking at suspensions of players for recreational drugs—drugs of abuse or recreation that have no performance-enhancement benefit, such as marijuana†—we find that almost all recreational drug use occurred among U.S.-born players. To date, *no* Dominican or Venezuelan baseball player has been suspended for recreational drug abuse.‡ Among the drug suspensions handed down to players from these countries, all were for performance enhancement. Among U.S. players, we found no relation between recreational drug use and economic background. Players from poorer neighborhoods were not more likely to use recreational drugs—just PEDs.

The stigma of using drugs may indeed differ across countries, and there may be a self-perpetuating effect. The greater proportion

---

* Specifically, we used census data on education, employment, and income for the city or MSA (metropolitan statistical area) in which the player was born in the decade in which he was born.

† One could claim that a few drugs of abuse might also enhance athletic performance. But most of the positive tests were for marijuana, which would not be performance-enhancing in almost any competition we could think of, the Nathan's Famous Fourth of July International Hot Dog Eating Contest notwithstanding.

‡ Kengshill Pujols, however, came close. A Dominican-born pitcher in the Dodgers organization, Pujols was already serving a 50-game suspension for PEDs in 2006 when he was arrested and found to be carrying 118 baggies of crack cocaine in his underwear. Charged with drug possession, he was released by the Dodgers and eventually convicted.

of Latin players who use such drugs, the less stigma it carries. The potential shame from such use, in theory, is lessened when "everyone else is doing it." And the more players who use banned substances, the more pressure it puts on others to cheat. If your class is being graded on a curve and you know many peers plan to cheat on the upcoming test, you're probably more tempted to cheat, too, just to keep up.

What begins as a result of economic incentives can grow into a self-reinforcing culture in which drug use is not only tolerated but expected. Soon athletes aren't cheating to get an edge; they're cheating simply to avoid falling behind (*see* professional cycling).

How big an effect do performance-enhancing drugs have? Answering this is difficult because it's hard to know the counterfactual: What would a player's performance have been had he not taken drugs? Although we don't claim to have a definitive answer, here's some empirical evidence. We looked at *all* major and minor league players and asked a simple question. Controlling for as many variables as possible—age, experience, height, weight, position, country of origin—how do the players who have been suspended for drug violations stack up against other players? What level within baseball's league system (from the Rookie league to Single-A, Double-A, Triple-A, and the Majors) did they achieve relative to other players with the same characteristics who didn't test positive?

As we discovered, suspended players were much more likely to achieve a higher level in the baseball league system. Specifically, an athlete suspended for using performance-enhancing drugs was 60 percent more likely to achieve the next level than an athlete who wasn't suspended. To gain some perspective on this, an additional year in a league increases the chances of making it to the next level by only 20 percent. So performance-enhancing drugs have three times the effect of an extra year of experience. For players born outside the United States, PEDs have an even greater effect, increasing the chance of a player moving up to the next level by almost 70 percent. In addition to the science, the data support the claim that steroids work.

To make this point another way, imagine a set of high school seniors of the same age, quality of education, and grade-point-average. Group A cheated on the SAT, and Group B didn't. If on average the students in Group A were admitted to higher-ranked colleges, you would reach the inexorable conclusion that cheating on the SAT had a benefit.

Could the reason that players caught using PEDs have achieved a higher level be simply that enforcement and drug testing are toughest at the highest levels in baseball? In other words, are baseball players cheating equally at all levels, but there's simply more scrutiny at the top?

In fact, just the opposite is true. There seems to be more vigilance at *lower* levels of baseball. The Major League Baseball Players Association, which has fought rigorous testing, represents only players in the Major Leagues. Without the full protection and advocacy of the union, minor leaguers have been subject to considerably more stringent testing—from a longer list of banned substances to more frequent testing. Not surprisingly, minor leaguers fail drug tests twice as frequently as do their Major League counterparts.

In light of what we've learned, it's also not surprising to find that, controlling for country of birth, age, and position, the *height and weight* of players caught using PEDs are much lower than those of other players. Shorter players and players who weigh less are far more likely to use a banned substance. Intuitively, this makes sense, and it also makes sense from an economic perspective. If performance-enhancing drugs improve power, for instance, it is the stature-challenged who would seem to benefit the most from such a boost. This evidence also seems to fly in the face of the "recovery from injury" claim. Unless recovery time is correlated to physical size—which no respected medical professional seems to claim—the explanation for the relationship between PED use and an athlete's frame is that players are taking it for power and mass, not to shorten recovery time from injuries.

In a lot of ways, athletes who use performance-enhancing drugs behave like borrowers in a financial market, leveraging an asset with risky debt. Plenty of market speculators invest more than 100 percent of their money by using leverage. If they have $10 and borrow $20 from the bank, they can invest $30, three times the original investment. The downside is that they have to pay the borrowed money back later (with interest), which exposes them to more risk: They could lose more than their original $10. When an athlete leverages his effort with PEDs, he is in effect taking his initial genetic inheritance—his body, his talents, his training, and his work ethic—and buying on credit. If his investment pays off, he'll exceed what he started with. If his marker is called and he can't cover his debts—that is, if he fails a drug test—he stands to lose not only his initial investment (his natural health) but his reputation and future as well. For some athletes, the risks and costs are too severe and outweigh the potential benefits. For others, including older athletes and players from poorer backgrounds, the potential benefits outweigh the costs.

Seen through an economic lens, many examples of drug cheating and deception—however deplorable and amoral we might find them—start to make sense. How many millions of Americans massage their income tax returns, calculating that the gain outweighs the potential penalties and the risk of detection? How many people better their prospects for a choice job by doctoring their credentials, reasoning that the worst thing that could happen would be to lose a job that wasn't theirs to begin with?

We spoke with Welington Dotel in the spring of 2010 as he was awaiting assignment with the Mariners. He expected to play for either the Clinton (Iowa) LumberKings of the Midwest League or the Everett (Washington) AquaSox, where he had spent the majority of the previous season. Sometimes he felt tantalizingly close to his dream of making the Majors. Other times his goal felt unfathomably distant. Money issues continued to loom large in his life. Having spent most of his signing bonus, he was now subsisting on a minor league salary of a few hundred bucks a week and living with a host family at every minor league outpost. He often phoned his family in the Dominican Republic with news of his baseball career, but they'd yet to see him play a game in the United States.

"Too expensive," he explained. "Maybe they come when I make the Majors!" It is not so hard to see why he might have felt compelled to bend the rules.

As we see it, fans are well within their rights to condemn baseball players for the steroids era. From Mark McGwire on down, the cheaters ultimately made the decision to cheat. But it bears pointing out that Major League Baseball and the players' association created a system that gave many players the choice between acting immorally and, at least from an economic perspective, acting irrationally.

DO ATHLETES REALLY MELT
WHEN ICED?

Does calling a time-out before a play actually work?

Plenty of field goal kickers have had rough games, but few have been so spectacularly bad that they've inspired an entire *Saturday Night Live* skit. It was during the 2005 NFL season that New York Giants kicker Jay Feely missed three potential game-winning field goals against the Seattle Seahawks. The Giants ended up losing 24–21. A few weeks later, *SNL* served up "The Long Ride Home: The Jay Feely Story." Feely, played by Dane Cook, is traveling aboard the team plane when the flight hits turbulence. He is asked to land the aircraft between two skyscrapers. Naturally, he drives it wide right.

Midway through the 2008 NFL season, it appeared as though Feely was prime for another round of mocking. By then, he was kicking for the other New York team, the Jets. With three seconds to go in an October road game against the Oakland Raiders, Feely trotted onto the field for a 52-yard attempt to send the contest into overtime. Feely went through his routine, struck the ball fairly cleanly, but doinked the kick off the goalpost. The Raiders crowd celebrated. Jets fans groaned.

But wait. Oakland's coach, Tom Cable, had called time-out

before the kick, a spasm of psychological warfare. Feely's attempt didn't count. After a brief pause, Feely tried again. This time, he drilled the ball through the uprights, sending the game into overtime. Afterward he explained that he happily welcomed the Raiders' time-out call. "I heard the whistle before I started, which is an advantage to the kicker. If you're going to do that, do that before he kicks," he said. "I can kick it down the middle, see what the wind does, and adjust. It helps the kicker tremendously."

The Raiders ended up winning the game in overtime—ironically, on a field goal of 57 yards, a heroic distance. But Feely was redeemed. And it was another bit of evidence that questioned the wisdom of "icing the kicker."

For decades, it's been an article of faith in the NFL that when Team A faces a pressure-infused field goal to tie or win a game, Team B calls a time-out to "make him think about it" or "plant seeds of doubt."

But does it work? Does icing the kicker increase the likelihood of a miss? Several years ago, two statisticians, Scott Berry and Craig Wood, considered every field goal attempt in the 2002 and 2003 NFL seasons, playoffs included. They paid special attention to "pressure kicks," which they defined as field goal attempts in the final three minutes of regulation or at any point in overtime that would have tied the game or given the kicking team the lead. Publishing their results in the journal *Chance,* Berry and Wood asserted that on pressure kicks between 40 and 55 yards, iced kickers were 10 percent less successful on average. (On shorter kicks, the effect was found to be negligible.) However, the statistical significance of the difference found—amounting to 4 kicks out of 39 attempts—has been questioned.

Nick Stamm of STATS, Inc., found that pressure kicks—defined as above except within the last two minutes of the game rather than the last three—in the NFL regular season from 1991 to 2004 showed an insignificant difference between iced and non-iced kicks. The conversion rate on iced kicks was 72 percent; for non-iced kicks, the rate was 71.7 percent. Stamm's work suggests that at best, icing the kicker does not diminish his chances of success.

We undertook our own study, using NFL data from 2001 through 2009 and using Stamm's standards for pressure kicks as well as some of our own—looking at kicks in the last two minutes, one minute, 30 seconds, and 15 seconds of the game. First, we looked at instances when the team on defense called a time-out right before the kick (icing the kicker) and compared that with instances when they didn't. We then controlled for the distance of the field goal attempt so that we could compare the same field goal from the same distance when one kicker has been iced and the other hasn't. Simply put, we found that icing made no difference whatsoever to the success of those kicks. NFL kickers being iced are successful from the same distance at *exactly the same rate* as kickers who are not iced. The following table shows the numbers.

NFL FIELD GOAL SUCCESS WHETHER OPPONENT CALLS A TIME-OUT OR NOT

PERCENTAGE OF KICKS MADE

Situation	All kicks	Iced	Not iced
Less than 2 minutes in 4th quarter or OT	76.2%	74.2%	77.6%
Less than 1 minute in 4th quarter or OT	75.5%	74.3%	76.4%
Less than 30 seconds in 4th quarter or OT	76.5%	76.0%	76.9%
Less than 15 seconds in 4th quarter or OT	76.4%	77.5%	75.4%

NFL data from 2001 to 2009, including playoffs.

In some instances, icing the kicker may exact a psychological price. In other instances, it may backfire, as it did with Jay Feely, giving the kicker the equivalent of a free dress rehearsal. In the vast majority of cases, the kick will be successful based simply on mechanics and how cleanly the ball is struck (and how well the ball is snapped and placed by the holder), not on whether the kicker had an extra 90 or so seconds to consider the weight of the occasion.

Former Tampa Bay Buccaneer kicker Matt Bryant in an interview with the *Tampa Tribune* summed it up this way: "I think when you're at this level, nothing like that should matter. If it does, you probably don't belong here."

Kickers aren't the only athletes opposing coaches try to ice. Take the waning seconds of an NBA game. A team is whistled for a foul. Just as the free throw shooter steps to the line, the opposing team reflexively calls a time-out to ice the shooter.

Just like icing the kicker in the NFL, it's a dubious strategy. We examined all free throws attempted in the last two minutes of the fourth quarter or overtime of all NBA games from 2006 to 2010 when the teams were within five points of each other—in other words, "pressure" free throws. When a time-out was called just before the free throws (icing), the shooter was successful an average of 76 percent of the time. When there was no time-out called, the free throw percentage was . . . 76 percent.

Next, we looked at only the first free throw attempt—the one directly after the time-out—and ignored the second since a player might have adjusted his shot after the first attempt or his nerves might have settled. There was, again, no difference. We even looked exclusively at situations in which the score was tied and thus a made free throw would have put a team in the lead. Again, there was no difference between shooters who were iced and those who weren't.

There might be valid reasons for a team to call a time-out before a high-pressure kick or free throw. The coach might want to devise a strategy to block the kick or set up a play in the event of a miss. The defensive team might want to create the appearance that it's doing *something* rather than standing by idly. They might want to ensure that the rights-paying television network has the opportunity to air an additional series of commercials—annoying the fans in the process. But they shouldn't expect the time-out to have any bearing on the subsequent play.

Icing doesn't freeze a player or heat him up. You might call it a lukewarm strategy.

THE MYTH OF THE HOT HAND

Do players and teams ride the wave
of momentum? Or are we (and they)
fooled into thinking they do?

In the sprawling clubhouse of the New York Mets, David Wright's locker is featured prominently, square in the middle of the room, near the front door. The symbolism is unmistakable. Wright isn't just a spectacular young third baseman and a handsome, genial guy born without the jerk chromosome. He is the face of the franchise. So it is that his locker is positioned in such a way that the media can always find him for a quote, teammates can observe how professionally he comports himself, and Mets employees can locate him when they need to corral him to meet with corporate sponsors or tape a promotional video or sign the cast of a season ticket holder's kid.

But early in the 2010 season there were few good vibes emanating from Wright's double-wide clubhouse stall. He was struggling at the plate: smacking nothing but air with his swings, grounding feebly into double plays, and taking pitches for called third strikes. After going hitless in four at-bats the previous night, Wright arrived for a late April game against the Chicago Cubs hitting .229, a full 78 points below his career average of .307. Only 11 of his 48 at-bats had yielded a hit. As with most baseball players, his mood moved in step with his batting average. He still answered questions from the media but did so in a clipped manner, staring ahead vacantly and shifting uneasily in his chair. He cut back on the sponsor grip-and-grins and the video promos and the other extra-baseball obligations that fall to a franchise's star player. He'd broken with routine and arrived at Citi Field early that day to take extra batting practice at the team's indoor cage, hoping "to get my swing back to where I want it to be."

Wright stressed that he wasn't "panicking or anything like that," and his remedies seemed sensible, especially given the range of cures that exist for slumping baseball players. Other batters have changed their diet, burned articles of clothing, and consulted team chaplains after a few hitless games. It is former Chicago Cub Mark Grace who's generally credited with coining the term *slump-buster* to describe a promiscuous, unattractive woman whom a struggling player "romances" in hopes of reversing his luck at the plate. Fans of the movie *Major League* will remember that the Pedro Cerrano character sought to extricate himself from a slump by sacrificing a chicken.

Still, Wright's slump—the perception of it, anyway—took on an aura of its own in the Mets' clubhouse. "Soon enough David will start hitting like everyone knows he can," said Jeff Francoeur, then the team's right fielder, who was batting over .300 at the time, 30 points higher than his career average. "Right now, until David gets going a little bit, the rest of us need to pick up the slack."

You might say that the Mets needed Wright to channel his inner Vinnie Johnson. Basketball fans will recall Johnson as the almost sphere-shaped sixth man from the Detroit Pistons' "Bad Boy" teams of the late 1980s and early 1990s. A shooting guard, Johnson probably was best known for his play in game five of the 1990 NBA finals. With the Pistons leading the series 3–1, the game tied, and 00.7 seconds on the clock, Johnson popped a 15-foot jumper to give Detroit the championship. A teammate suggested that

Johnson be called 007, a nod to the time remaining when he hit his winning shot as well as to James Bond's sharpshooting.

The nickname didn't stick, though, mostly because Johnson already answered to one of the all-time great handles in sports. He was, of course, "the Microwave." The nickname was conceived by Danny Ainge of the rival Boston Celtics, who, like so many, marveled at Johnson's ability "to get hot in a hurry." Johnson was a classic streak shooter—a "rhythm guy," as they say in NBA-speak—capable of both spectacular and spectacularly bad marksmanship. Like all athletes, he endured slumps. But it seemed that once he dialed in his shot, his awkwardly released jumper found the bottom of the net with brutal accuracy. "When he came in and hit his first shot, everyone knew: *Look out*," Chuck Daly, the late coach of the Bad Boys, once said. "I've never seen a player who used momentum the way Vinnie did."

Momentum is such a vital component of sports that it's taken on the qualities of a tangible object. Teams and athletes have it. They own it. They ride it. They take it into halftime, into the series, into the postseason. They try like hell not to give it back or lose it. A slumping player like David Wright needs to get his momentum back—and he did, by the way, boosting his season average to .314 by the time he played in the 2010 All-Star Game. A player like Vinnie the Microwave Johnson was thought to use momentum so effectively that it came to characterize his 14-year career.

But what if we told you that momentum doesn't exist in sports?

■ ■ ■

First, let's be clear. Indisputably, *streaks* occur in sports. In any league, in any sport, and at any level, teams and athletes perform well and perform poorly, sometimes for significant stretches of time. As of this writing, the Pittsburgh Pirates haven't had a winning baseball season since 1992. That is a dismal run of consecutive sub-.500 seasons; in fact, it's the longest in the history of major American team sports. When LeBron James left Cleveland in the summer of 2010, we were told (and told and told) that no local team had won a championship since 1964. On a more successful note, at this writing, the girls' tennis team at Walton High School outside Atlanta hasn't lost since 2004, a run of 133 straight matches. Speaking of tennis, Roger Federer has won at least one Major tennis championship every year since 2003. All of these are streaks, no doubt about it.

The real question is whether those streaks *predict* future performance. If you make your last few field goals or putts or send fastballs over the outfield fence, is your next attempt more likely to be successful? Does the team or player currently "riding the wave" fare better or worse in the next game than one who is not? Does recent performance directly influence immediate future performance? Or are streaks nothing more than random chance, the outcome of luck, predictive of nothing?

As the minor legend of Vinnie Johnson suggests, momentum is probably cited most often in the NBA. Some players, we're told, cultivate a hot hand, and others cool off. Teams, we're told, carry momentum and come into a game "on a roll." This isn't a new observation. Professors of psychology and behavioral economics Thomas Gilovich and Robert Vallone, then from Cornell, and the late Amos Tversky from Stanford studied basketball momentum and the hot hand phenomenon a generation ago. They followed the field goal attempts of nine Philadelphia 76ers during the 1980–1981 season and found no evidence of momentum. Field goal success, they reckoned, is largely independent of past success on recent attempts. Successful shot making—or missing—had no bearing on a player's next attempt.

Of course, field goal success may be affected by what the defense is doing. A player who has hit several shots in a row may be guarded more vigilantly, which might make his success rate on the next attempt lower. Likewise, a player who has missed his recent shots might face a more lax defense, which could mean a greater likelihood of success on his next attempt. To avoid these potential distortions of a hot or cold hand, the professors also looked at free throw shooting, which involves no defense or adjustments. Their subjects were nine players from the Boston Celtics during the 1980–1981 and 1981–1982 seasons. Again they found no momentum or evidence of a hot hand. What players did on their previous free throws didn't affect what they did on the next free throw.

The psychologists then looked across games and saw that being "hot" or "cold" one night did not predict performance the next night. It wasn't necessarily that players who were "unconscious" one game automatically "came back to earth" the next game or that players who'd lost their touch in one game necessarily regained it the next night. Rather, there was simply no evidence that the streaks had any carryover effect; they simply were not predictive of future performance.

In addition to looking at NBA players, the researchers conducted an experiment using the varsity basketball players of the men's and women's teams at Cornell. They had the players shoot successive free throws and field goals from the exact same spot on the floor, facing no defensive, crowd, or game pressure. Once again, they found no evidence of the hot hand effect. Players who hit several shots or free throws in a row were no more likely to hit the next shot than were players who had missed several shots in a row.

Here's the interesting part. The players themselves—both in the NBA and at Cornell—firmly *believed* in the hot hand effect. They *felt* hot or cold, as though the result of the previous shot would go a long way toward determining the result of the next one. Considering how often coaches instruct their players to "feed the hot hand," it's clear that many of them believe in the phenomenon of momentum. And as fans, most of us do, too. Before the professors began looking at basketball players' stats, they surveyed fans and found that 91 percent agreed that a player has "a better chance of making a shot after having just *made* his last two or three shots than he does after having just *missed* his last two or three shots." In fact, the fans estimated that his chances were 20 percent greater if he had just made his last two or three shots. Even for free throws, 68 percent of the surveyed fans agreed that a player has "a better chance of making his second shot after *making* his first shot than after *missing* his first shot." A full 84 percent of fans believed that "it is important to pass the ball to someone who has just made several shots in a row."

Because the researchers' data was so thoroughly at odds with perceptions in the sports world, they and other researchers refined the study further. Maybe shots taken within one minute of each other would exhibit more persistence. But that didn't turn out to be the case. They tried replicating their findings using more data on more players over more seasons. Still, the results remained unchanged. Others looked at the results of the Three-Point Shootout held during the NBA's All-Star Weekend, in which the most accurate shooters on the planet—absent defense, in a controlled environment—compete in a straightforward contest. Again, as often as announcers declared, "He's on fire!" there was no evidence of momentum.

There was one academic study that, initially anyway, *did* find evidence of a hot hand. Irony of ironies, the results were driven largely by . . . the shooting of Vinnie Johnson. The key piece of supporting evidence, though, was the Microwave's run of seven consecutive baskets in the fifth game of the 1988 NBA finals. Unfortunately, that seven-out-of-seven streak never happened. The data had been miscoded. (He missed his fourth shot, though a teammate tipped in the miss.) Once the data were corrected, Johnson—again, the player most notorious for shooting in spurts—was shown to be no more or less streaky than any other player, no more or less likely to make a shot after a hit as he was after a miss. But the researchers did show that Johnson and his teammates *thought* he was a streak shooter. He tended to shoot more after making a basket and was fed the ball more frequently after each make. The problem was, he wasn't more likely to score.

More recently, John Huizinga and Sandy Weil, who also investigated the value of blocked shots (see "The Value of a Blocked Shot"), updated the hot hand study by looking at all NBA games between 2002 and 2008. They, too, found no evidence of any hot hand effect. However, they did find something else. Despite there being no greater likelihood of accuracy, shooters making their last several attempts act as if a hot hand exists. After making a shot, they take harder shots—and shoot about 3.5 percentage points below their normal field goal percentage. They also shoot 16 percent sooner than they do after a missed jump shot and are almost 10 percent more likely to take their team's next shot if they made their last shot than if they missed it. (Both of these effects are much stronger for point guards and swingmen, which stands to reason: No one talks about a "hot hand" in conjunction with dunks, layups, and other short-range shots.) The authors concluded that if everyone on the team behaved this way—shooting more frequently and taking more difficult shots after a previous make than after a previous miss—it could ultimately cost a team 4.5 wins per season on average.

Okay, so momentum doesn't exist on the level of the individual, but what about at the team level? Does momentum exist for NBA teams? We considered about 3,500 NBA games between 2005 and 2009, examining the play-by-play data and paying special attention to scoring runs. One can define scoring runs in any number of ways; we chose to look at teams that scored at least six unanswered points in the previous minute and called them hot (or their opponents cold). We then asked what happens over the next minute in those games. Did the hot team continue to remain hot by increasing its lead (or decreasing its deficit) against the cold team?

In a word, no. In fact, we found the opposite. If a team scores six or more unanswered points in the previous minute, it will on average be outscored by its opponent (by 0.31 points) over the next minute. This implies that there isn't merely an absence of momentum; there is a *reversal*. A team that gets hot is more likely to do worse, not better.

But perhaps a minute was too short a time frame to consider momentum, so we looked over the next two, five, and ten minutes. But we found the same effect: reversals of fortune, not evidence of momentum. The following chart sums up our results.

In every instance—no matter how far back we defined hot or cold teams and how far forward we looked—we found strong evidence of reversals rather than momentum. Hot teams tend to

WITHIN-GAME MOMENTUM IN THE NBA:
POINT DIFFERENTIAL OF "HOT" TEAM
OVER SUBSEQUENT MINUTES

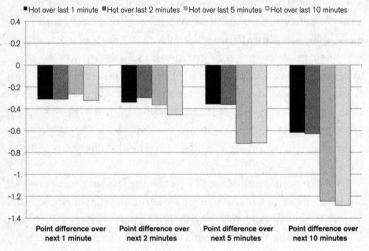

■ Hot over last 1 minute ■ Hot over last 2 minutes ▨ Hot over last 5 minutes ▢ Hot over last 10 minutes

"Hot" = scoring 6 (9, 14, 20) or more unanswered points
than the opponent over the last 1 (2, 5, 10) minutes.

get outscored after going on a run; cold teams tend to catch up. When he hosted ESPN's *SportsCenter,* Steve Levy had a catch-phrase: "It's the NBA; everyone makes a run." Turns out, he was absolutely right.

This, of course, could be for a variety of reasons. Maybe streaking teams expend more energy when they make a run and then get fatigued. Perhaps after a big run coaches are more likely to send in inferior bench players. Or perhaps players exert less effort after they've built a comfortable lead, or opponents exert more effort when behind. Whatever the reason for the reversals, the evidence supporting momentum is simply not there.

Next, we looked at comebacks. A team is down by, say, ten points in the waning minutes of a game and stages a furious come-back to tie the game and send it into overtime. Does that team have a better chance of winning in OT? The answer, we've found, is no. Its chance of winning in overtime is no different from that of the team that gave up the lead (or, for that matter, than it is for two teams that were neck and neck the entire game before heading to OT). We found no evidence that teams that are on winning streaks of two, three, four, five, six, seven, eight, nine, or even ten games have any better chance of winning the next game. (The same is true for teams on long losing streaks.)

Even the postseason seems immune to momentum. Often we hear how important it is that teams get hot or sustain momentum going into the playoffs. We find no evidence for that. Controlling for a team's overall regular season record, we find that a team entering the playoffs on a winning streak—even as much as ten games—does no better than a team entering the postseason on a losing streak. (Sure enough, as we were studying the data, the Boston Celtics were marching to the 2010 NBA finals, having lost seven of their last ten regular season games. Their opponents, the Lakers, had lost six of their nine final regular season games.)

Nor is the absence of momentum unique to the NBA. In baseball, hitting streaks seem to be no more predictive of future success than slumps are. Batting averages of players are just as likely to be higher after cold streaks as they are after hot streaks. The same thing goes for pitchers. We found no evidence that starters get into "a groove" that enables us to predict future success. Researchers *have* found the existence of momentum in two niche sports, bowling and billiards, but those sports (games, really) rely on a repetitive motion and take place in the same physical space. That's a lot different from sinking jump shots in the face of a defense or hitting 95-mph fastballs.

■ ■ ■

Why do we attribute so much importance to "sports momentum" when it's mostly fiction? Psychology offers an explanation. People tend to ascribe patterns to events. We don't like mystery. We want to be able to explain what we're seeing. Randomness and luck resist explanation. We're uneasy concluding that "stuff happens" even when it might be the best explanation.

What's more, many of us don't have a firm grasp of the laws of chance. A classic example: On the first day of class, a math professor asks his students to go home, flip a coin 200 times, and record the sequence of heads and tails. He then warns, "Don't fake the data, because I'll know." Invariably some students choose to fake flipping the coin and make up the results. The professor then amazes the class by identifying the fakers. How? Because those faking the data will record lots of alternations between heads and tails and include no long streaks of one or the other in the erroneous belief that this looks "more random." Their sequence will resemble this: HTHTHHTHTTHTHT.

But in a truly random sequence of 200 coin tosses, a run of six or seven straight heads or tails is extremely likely: HTTTTTHHTTTHHHHHHH.

Counterintuitive? Most of us think the probability of getting six heads or tails in a row is really remote. That's true if we flip the coin only 6 times, but it's not true if we flip it 200 times. The chances of flipping 10 heads in a row when you flip the coin only 10 times are very low, about 1 in 1,024. Flip the coin 710 times and the chances of seeing at least one run of 10 straight heads is 50 percent, or one in two. Flipping the coin 5,000 times? We'd see at least one string of 10 in a row 99.3 percent of the time. At 10,000 times, it's virtually a lock (99.99 percent) that we'll see at least one run of 10 heads in a row.

In 1953, the psychologist Alphonse Chapanis of Johns Hopkins documented how badly human subjects understand randomness by asking them to write down long sequences of random numbers (0 through 9). He found that almost no one chose to use repetitive pairs such as 222 and 333. Subjects instead tended to alternate digits and avoid repetition. In short, they couldn't create random sequences. This bias can be gamed or taken advantage of. State lotteries, for instance, have an overwhelming number of tickets with alternating numbers and very few with repetitive digits. Since most lottery pots are split among the winners, your chances of having the pot all to yourself are greater if you pick 22222 versus 65821. Nobody picks 22222. But why not? The lottery is random. Drawing 22222 is just as likely as drawing 65821. You have to get all five digits correct and in the same order in either case. But people don't see it that way. (Imagine if the Powerball numbers actually came out 22222. Most people's first thought would be that something was suspicious.)

The same thing is true with flipping a coin. If you get ten heads in a row, what's the likelihood that the next flip will be heads? Don't be fooled—it's 50 percent, the same as it is on any single coin flip. Most people think the chances of getting heads will actually be *lower* than 50 percent—the opposite of momentum. They know they should see roughly the same number of heads as tails (50–50), so they feel that if they've seen ten heads in a row, they're *due* for a tails. A tails has to emerge to balance things out. But it doesn't. There is no law of averages. If the process is random, there's no predictability. This is also what drives the "gambler's fallacy." Gamblers on losing streaks erroneously believe they're due and keep gambling, thinking that their luck has to balance out. But if the whole thing is random, you aren't due for anything. Your chances haven't changed at all.

The casinos, of course, are happy to exploit this failure to understand randomness. Some of them even post the recent results of the roulette wheel spins, hoping to dupe gamblers: "Hey, it's landed on odd five straight times. We're due for even!"

■ ■ ■

Now the contradiction between the strong belief in the hot hand or momentum in sports and the lack of actual evidence starts to make sense. A basketball player who shoots 50 percent will not miss an attempt and then make an attempt. A batter may hit .300, but it's only an average. It doesn't mean that he'll get three hits in *every* ten at-bats. He might go 0 for 10 and then 6 for 10. Over the 600 at-bats throughout a season, however, he *probably will* get 180 hits. The larger a sample, the more accurately it represents reality.

Kobe Bryant shoots free throws much better than Shaquille O'Neal does. For their respective careers, Kobe hits about 84 percent from the line, and Shaq only 53 percent. Take a sequence of only five shots, however, and it's entirely possible they'll shoot comparably. It's even reasonably possible that Shaq will outshoot Kobe. In fact, the chances are about 22 percent; that means if Shaq and Kobe staged a five-shot free throw shooting contest, about one out of every five times Shaq would do at least as well as Kobe and might even beat him. Over ten attempts, it's less likely. Over 100, it's remote.

What does this mean with regard to David Wright's hitting slump? A career .307 hitter, Wright expects to get a hit 30 percent of the time. Three weeks into the season, after getting a hit only 23 percent of the time, his performance is perfectly consistent with his .307 average. The same would be true on the other side. He could have hit .400 the first few weeks, and fans would be ready to declare him the first player since Ted Williams to bat .400 for a season. Yet over a short period, a .307 career hitter batting .400 is perfectly consistent with random chance, too. Some athletes get this better than others and try to avoid "getting too high or too low." Wright's former teammate Jeff Francoeur performs a self-assessment on hitting every 50 at-bats. But even that is woefully narrow.

■ ■ ■

Being fooled by chance can create seemingly unbelievable statistics. Consider the following, all of which are true. Over the last decade, *in every single* MLB season:

- *At least one National League pitcher has had a longer hitting streak than a starting All-Star (nonpitcher).*
- *At least one National League pitcher has had a longer hitting streak than a designated hitter in the American League.*
- *At least one batter hitting under .225 for the season has put together a hitting streak longer than that of a player hitting over .300.*
- *At least one player who finished the season hitting over .300 has had at least one six-game or longer hitless streak.*

These stats, surprising as they might seem on their face, hold up every year. Pitchers are not supposed to hit better than position players, much less all-stars. Players hitting under .225 aren't supposed to have longer hitting streaks than .300 hitters. The best batters aren't supposed to go six games—25 or so at-bats—without a hit, and on average, they don't. But in isolated cases it happens, and it's perfectly consistent with random chance.

We search for an explanation, but the true explanation is simple: Luck or chance or randomness causes streaks among even the best and worst players. It has nothing to do with momentum. When we consider the bigger picture and the larger numbers of players in Major League Baseball, this starts to make sense. How likely is it that Tim Lincecum, the star pitcher for the San Francisco Giants, will outhit the mighty Albert Pujols over any stretch of the season? Not very. How likely is it that *any* pitcher will outhit Pujols over a two-week stretch? More likely. How likely is it that *at least one* pitcher will outhit *at least one* all-star position player over those weeks? Very, very likely. The larger the sample, the more you can find at least one seemingly unlikely example.

If you were predicting the likelihood of an MLB player getting a hit in his next at-bat, which of the following do you think would be the best predictor?

a) His batting average over the last five plate appearances
b) His batting average over the last five games
c) His batting average over the last month
d) His batting average over the season so far
e) His batting average over the previous two seasons

Most people are tempted to select (a), on the grounds that it is the most recent and therefore the most relevant number: He's streaking and will continue riding the wave. Or, he's slumping and still mired. But to pick (a) is to be fooled by randomness, tricked into thinking there's momentum.

We looked at all MLB hitters over the last decade and tried predicting the outcome of their next at-bats by using each of the five choices above. It turns out (a) is the worst predictor. Why? Because it has the smallest sample size. Choice (b) was the next worse, then (c), and then (d). The best answer was (e), the choice with the largest sample size.

The same thing is true at the team level. Heading into the postseason—and barring the unusual, such as a recent horrific injury to a star—which of the following is a better predictor of playoff success?

a) The team's performance in its most recent game
b) The team's performance in the last week before the playoffs
c) The team's performance in the last month before the playoffs
d) The team's regular season performance

Momentum would lead one to think that it's (a) or (b) and, to a lesser extent, (c), yet those are actually the worst predictors. In *every single sport* (MLB, NBA, NHL, NFL, European soccer) we studied, we found (d) to be the best predictor of postseason or tournament success. The true quality of teams can be measured best in large samples. Small samples are more dominated by randomness and therefore are inherently unreliable.

■ ■ ■

Nor is this unique to sports. In the investment management industry, investors often "chase short-term returns," flocking toward mutual funds that had a good quarter or year and fleeing from funds that didn't. They ascribe success on the basis of a small sample of data. But as with the hot hand in sports, it turns out that one quarter or even one year of a fund's performance has no special predictive power for the next year's performance in the mutual fund industry. In fact, one year of performance for almost any fund is dominated by luck, not skill. Yet people usually don't see it that way. Entire businesses have been built on selling short-term performance measures to investors to help them identify the best funds, and funds aggressively market their recent strong performance to investors (and hide or bury bad performance when they can). But the reality is that every year the top 10 percent of funds are just as likely to be among the bottom 10 percent of funds the next year. It's just pretty much random from year to year.

Sports gamblers, too, are fooled by momentum. Colin Camerer, a Caltech professor of behavioral economics, found that winning and losing streaks affected point spreads. Bets placed on teams with winning streaks were more likely to lose, and bets placed on teams with losing streaks were more likely to pay off. In other words, gamblers systematically overvalued teams with winning streaks and undervalued those with losing streaks.

Just as an astute investor can take advantage of these misperceptions with potentially big gains, so can a savvy coach and player (and sports gambler). If the majority overvalues the recent winners and undervalues the recent losers, do the opposite.

The only problem is convincing people to go against their (and everyone else's) intuition. After the initial study asserting the fallacy of the hot hand in basketball, Red Auerbach, the revered Hall of Fame coach and then president of the Boston Celtics, was presented with the findings. Auerbach rolled his eyes and waved the air with his hand. "Who is this guy? So he makes a study. I couldn't care less." Bob Knight, the volatile and decorated college coach, was similarly dismissive: "There are so many variables involved in shooting the basketball that a paper like this doesn't really mean anything." Amos Tversky, the famous psychologist and pioneering scholar who initiated the original research on momentum and the myth of the hot hand, once put it this way: "I've been in a thousand arguments over this topic. I've won them all, and I've convinced no one."

DAMNED STATISTICS

Why "four out of his last five"
almost surely means four of six

"There are three kinds of lies: lies, damned lies,
and statistics."—Mark Twain

At some point it became almost cartoonish, as though he wasn't shooting the basketball so much as simply redirecting his teammates' passes into the hoop. In the first half of the second game of the 2010 NBA finals, Ray Allen, the Boston Celtics' veteran guard was . . . well, the usual clichés—"on fire," "unconscious," "in the zone"—didn't do it justice. Shooting with ruthless accuracy, Allen drained seven three-pointers, most of them bypassing the rim and simply finding the bottom of the net. Swish. Swish. Swish-swish-swish. In all, he scored 27 points in the first half. Celtics reserve Nate Robinson giddily anointed Allen "the best shooter in the history of the NBA."

As Allen fired away, the commentators unleashed a similarly furious barrage of stats, confirmed by the graphics on the screen. The shooting was cast in the most glowing terms possible. Allen, viewers were told at one point, had made his last four shots. When he missed a two-pointer (turns out he was only three for nine on two-point attempts), the stats suddenly focused only on the three-pointers.

It was inevitable that Allen would cool off. And he did in the second half, making only one three-pointer, although his eight treys for the game became a new NBA finals record and his 32 total points enabled Boston to beat the Los Angeles Lakers 103–94. But he *really* cooled off in his next game. This time he was ruthless in his *in*accuracy, missing all 13 of his shots, including eight three-point attempts, as Boston lost 91–84. As Allen clanged shot after shot, the commentators were quick to note this whiplash-inducing reversal of fortune, framing it in the most damning terms possible. At one point viewers were told that between the two games, Allen had missed 17 straight attempts.

Inasmuch as sports fans are tricked by randomness, the media share in the blame. Statistics and data are the forensic evidence of sports, but like all pieces of evidence, they can be mishandled and tampered with. We are bombarded by stats when we watch games, but the data are chosen selectively and often focus on small samples and short-term numbers. When we're told that a player has reached base in "four of his last five at-bats," we should assume right away that it's four of his last six. Otherwise, rest assured, we'd have been told that the streak was five out of six. Clearly, a team that "has lost three in a row" has dropped only three of its last four—and possibly three of five or three of six or . . . otherwise it would have been reported as a four-game losing streak.

■ ■ ■

Those of us in the sports media have an interest in selling the most extreme scenario. Collectively, they (we?) pick and choose data accordingly. Take, for instance, a September 15, 2009, game between the New York Yankees and the Toronto Blue Jays, a showdown between Alex Rodriguez and Roy Halladay, arguably the league's best hitter and best pitcher at the time. The Yankees broadcasters might have framed the encounter along the following lines, using the most positive statistics at their disposal:

> *Rodriguez steps to the plate. He's hitting .357 against Halladay this season, including five hits in his last 12 at-bats against the big righty, a .412 clip. Over his last 11 games, A-Rod is hitting .436. Remember that as trade rumors swirl, Halladay has lost 4 of his last 5 starts and 11 of his last 15.*

Upon receiving this information, it sounds almost like a foregone conclusion that A-Rod is going to crush the ball. In fact, one almost feels pity for Halladay. It's as though he should have taken the mound wearing a helmet and protective covering.

Now listen to how the Toronto broadcasters might have addressed the showdown, using the best available statistics to make their case:

> *Halladay comes in having pitched two straight complete games. Over those 18 innings, he struck out 18 men and gave up only four earned runs, a 2.00 ERA. Meanwhile, A-Rod is hitless in his last six at-bats against Halladay. Among all opposing teams, Rodriguez has his lowest average—and strikes out the most—against the Blue Jays.*

After hearing this we'd be surprised if Rodriguez made contact with a Halladay pitch, much less reached base.

Both renderings would have been perfectly accurate. Both sets of statistics are true. Yet they paint radically different pictures. Incidentally, in that Yankees–Blue Jays game, Halladay pitched six innings, allowed two earned runs, and got the win; Rodriguez was one for three with a double against Halladay—pretty much what a neutral observer, ignoring the noise and looking at as much data as possible, would have predicted.

■ ■ ■

Teams are complicit in this selectivity, too. Check the scoreboard next time you're at a baseball game. Had you attended a White Sox–Tigers game at U.S. Cellular Field in the summer of 2010, you could have learned that Chicago's outfielder Carlos Quentin was "hitting .371 over his last nine games." Although this was impressive and meant to convey a hot streak, it told us . . . what exactly? Not much, not with a sample size that small. If the White Sox were attempting to predict the outcome of Quentin's next at-bat, they would have provided a more meaningful statistic, using a larger data set. But noting that Quentin was "351 for 1,420 (.247) for his career" doesn't quite stir up passion.

When Nate Robinson declared Ray Allen the best shooter in the annals of the NBA, he may have been right, but not because Allen had one torrid shooting half. Otherwise, you could just as easily make the case that based on the following night's game, Allen was the worst shooter in NBA history. Robinson's more convincing evidence would have been this: For his NBA career Allen has taken more than 6,000 three-point attempts and made roughly 40 percent of them.

Those two games of extremes during the 2010 NBA finals? Unsexy as it might have been to use the largest available data set and note Allen's career average, it would have helped the viewers. Between the two games, he was 8 of 19, or 42 percent, on three-point attempts, conforming almost exactly with his career mark.

ARE THE CHICAGO CUBS CURSED?

If not, then why are the Cubs so futile?

The ball collided with the bat of Luis Castillo and made a hollow *thwock,* the tip-off that it hadn't been hit cleanly. It wafted into the autumn night sky and descended between the foul pole and the third-base line at Wrigley Field. The Cubs left fielder, Moises Alou, ambled over, tracking the ball.

It was a foregone conclusion that Alou would make the catch, completing the second out of the eighth inning in this, the sixth game of the 2003 National League Championship Series. Ahead of the Florida Marlins 3–0 this night and leading in the best-of-seven series three games to two, the Cubs would then be just four outs from reaching the World Series for the first time since 1945. Already champagne was nesting on ice in the Cubs' clubhouse. The Marlins' team president had just called his wife to tell her there was no need to come to Chicago because there would be no game seven. Alou, a capable fielder and a veteran of several all-star games, positioned himself under the ball. The crowd, already rock-concert-loud, thickened its roar. Alou extended his left arm, yelled "Got it, got it," jumped up alongside the stands, unfurled his glove, and . . .

You probably know the rest of the story. A 26-year-old consultant had managed to score a sweet ticket for this game: aisle 4, row 8, seat 113, the first row before the field. A mishandled nacho and the cheese would have landed in the dirt of foul territory. For Steve Bartman, this was about as close to nirvana as he could get. A Chicago native, Bartman was the kind of long-suffering citizen of Cubs Nation whose moods moved in accordance with the team's fortunes. Bartman had recently graduated from Notre Dame but had returned home, yes, because of his job and family but also because of the proximity to his beloved baseball team. His level of fandom was such that despite his prime seat, he still listened to a radio broadcast of the game on headphones as he watched.

When Castillo's foul ball traced an arc and began its downward flight, Bartman rose to catch it, a reaction almost as instinctive as withdrawing one's hand from a hot flame. A souvenir foul ball? What better way to garnish a magical night. In less time than it will take you to read this sentence, Bartman's life—to say nothing of his magical night—was turned on its head. In his zeal to catch the ball, he interfered with Alou and knocked the ball away. After realizing an out had been lost, Alou popped away as if bitten by a snake. He shot Bartman a death stare and, in a gesture unbecoming a 37-year-old man, slapped his glove in the manner of a Little Leaguer throwing a tantrum. "Alou, he is livid with a fan," intoned the television broadcaster. Mark Prior, the Cubs pitcher, turned to left field and also stared darts into Bartman.

Given new life, Castillo walked. It was around that time that Bartman was escorted from his seat by security. "It's for your own safety," he was told. Even then, he was heckled and cursed and doused in beer. Bartman buried his face in his sweatshirt as if doing a perp walk through Wrigley—the ballpark, incidentally, nicknamed "The Friendly Confines."

It was a good thing security arrived when it did. Castillo's walk catalyzed an eight-run rally. There were wild pitches and cheap hits and an error by the Cubs shortstop, Alex Gonzalez, on what should have been an inning-ending double play. The Marlins won the game. By then, Hollywood production companies were already angling for movie rights. According to the next day's *Variety, Fan Interference,* starring Kevin James, would tell the story of "a [fan] who screws up an easy out, and then has to deal with the ramifications." Thanks to the speed and power of the Internet, Bartman's identity was revealed by morning. At his office at Hewitt Associates, a management consulting firm in the North Shore suburbs, his voice mail was clogged with profane messages. Bartman released a statement, stating that he was "sorry from the bottom of this Cubs fan's broken heart."

No matter. The following night, the Cubs would lose the decisive seventh game to the Marlins, who would go on to win the World Series. And poor Steve Bartman would take his rightful place alongside Mrs. O'Leary and her cow among the city's bêtes noires. With Halloween a few weeks away, Steve Bartman masks began appearing at parties, outnumbering witches and ghosts and Osama bin Laden costumes by a healthy margin. A comic at the famed Second City comedy theater soon did a routine dressed as Bartman, bumping into a fireman as he tries to catch a baby from a burning building. A local radio station began playing a song, "Go Blame It on Steve Bartman." Then there were the T-shirts spoofing the MasterCard "priceless" commercials:

> Tickets to a Cubs game: $200
> Chicago Cubs hat: $20
> 1987 Walkman: $10
> F—ing up your team's chances of winning the World Series: Priceless

Law and Order began taping an episode about a "foul ball guy" who deprived his team of a victory and was subsequently found murdered in a bar. Interviewed on the local news, Dan May, a local law student, explained that if he were to cross paths with Bartman, "I wouldn't shoot him. But I'd break his knees." Richly, Rod Blagojevich, then Illinois's governor, stated that Bartman "better join the witness protection program." He added that if Bartman had committed a crime, "he won't get a pardon from this governor." Meanwhile, Florida's governor at the time, Jeb

Bush, jokingly offered Bartman asylum. Bartman went into hiding and declined to give interviews or make appearances, including an invitation to attend the Super Bowl. There were rumors that he underwent plastic surgery. At the time of this writing, he has yet to resurface in public.

Bartman didn't catch the ball that night. It squirted away from him and eventually was recovered by a Chicago lawyer who sold the ball at auction to Harry Caray's Restaurant Group. The winning bid? $113,824.16. That off-season, the ball was destroyed in a ceremony that drew more than a thousand Cubs fans and Chicago celebrities on the order of Smashing Pumpkins lead singer Billy Corgan, *Caddyshack* and *Vacation* director Harold Ramis, and Caray's widow, Dutchie. The detonation was overseen by Michael Lantieri, an Oscar-winning Hollywood special effects savant. (Disregarding suggestions to use Caray's thick glasses to ignite the fire that would melt the ball, Lantieri utilized a no-smoke detonation device.) Fans stood outside the tent—some wielding signs reading "Death to Bartman"—chanting "The ball is dead. The ball is dead." The steam from the ball was gathered, distilled, and used to prepare pasta sauce for Caray's restaurant. Really.

■ ■ ■

If all this sounds a bit—how to put it?—*extreme,* to many fans "Bartman's blunder" confirmed what they had already believed for so long: The Cubs are simply doomed, a star-crossed franchise that has done something to anger the fates. A curse once ascribed to a vengeful billy goat now had a human face, one wrapped in a terminally uncool set of headphones.

Maybe the misbegotten fans—and, for the sake of full disclosure, we count ourselves among the legion—were on to something. The last time the Cubs won the World Series was 1908—the longest championship drought in all major North American professional sports. For the sake of perspective: Teddy Roosevelt was president, the concept of a world war was yet to be conceived, and the horse and buggy was more common than the automobile. It was Jack Brickhouse, the longtime Cubs play-by-play announcer, who noted, "Everyone is entitled to a bad century."

When Bartman made that awkward attempt to catch the ball, it was as if karma, suddenly awakened, reminded us that the Cubs and success don't mingle. After Castillo's walk, a parade of miscues ensued, hits dropped between fielders and the fortunes of the team did a pirouette. Proof of a higher power at work. The Cubs were—indeed, *are*—simply the unluckiest franchise in all sports. In a word: cursed.

Right?

Well, first, let's define luck. Webster's weighs in with this: "a force that brings fortune or adversity . . . something that brings extreme outcomes that are unexpected." So how "unlucky" are the Cubs? For example, how unlucky is it that a team would play for an entire century without winning the Big Prize once? In a league with 30 teams, as there currently are in baseball, assuming that each team has an equal chance at winning, the odds that one would play for 100 seasons and never win the big prize are about 3.5 percent, or 1 in 30. Unlikely but hardly impossible.

If you're looking for a real outlier based on strict odds, think back to the New York Yankees' achievement of winning 27 of the last 100 World Series crowns. (There was no World Series in the strike year of 1994.) The chances of that? One in 32 billion, or 1:32,216,000,000.

No one attributes the Yankees' remarkable success to luck. This kind of luck simply doesn't happen. The Yankees have just been really good, employing some of the most gilded players—Ruth, DiMaggio, Mantle, Jeter—scouting and developing talent, and hiring expert coaches. (Yes, at least in recent years, they've also spent a boatload of money.) The Yankees may be a lot of things, but no one, at least outside Boston, is arguing that the franchise is *lucky*. So why should we ascribe the Cubs' remarkable lack of success to luck? That is, why are we quick to embrace luck for the Cubs' failure and reluctant to do so for the Yankees' success?

Luck is something we can't *explain*. It is often attributed to things we don't want to explain. Psychologists have found that people too often attribute success to skill and failure to luck, a bias called self-attribution. We brag about the three stocks we bought that hit it big but dismiss as bad luck the seven that plummeted. We applaud our quick reflexes and driving skills when avoiding a gaping pothole, but when we hit it squarely, we curse the weather, other drivers, and the city (everyone but ourselves). In many aspects of life, we are quick to claim success and reluctant to admit failure. We do the same thing for our favorite team.

A curse, or bad luck, is an easy way out. When attributing failure to luck, you need search no further for an answer. To borrow a favorite phrase from baseball clubhouses, "It is what it is." Bad luck has the beautiful, comforting quality of getting you off the hook. Failure is unavoidable if it's due to luck. It was out of your control, and there is nothing you could have done or should have done to change it.

Is the real explanation for the Cubs' futility being masked by the convenience of a so-called curse? And if so, is there perhaps something that can be done to change the franchise's fortunes rather than sacrificing fumbled foul balls to the baseball gods?

To answer these questions, we need to measure something that is inherently immeasurable: luck. Although we can't directly measure luck itself, we can infer from data where luck has had its influence.

Consider again what it means to be unlucky. The term implies a certain randomness or a lack of control; in other words, outcome that isn't commensurate with ability. A team that consistently wins its division yet never wins the World Series? That's unlucky. A franchise that consistently finishes second in its division despite having a great team and record, perhaps because the ball didn't bounce its way a few times? Or it happened to be in a division with a mighty powerhouse such as the Yankees? That's unlucky. A team that loses a lot of close games may be unlucky. A team performing well on the field in every measurable way but failing to win as many games as it should? Again, unlucky.

That said, how much of the Cubs' futility can be attributed to bad luck? To win a World Series, you have to get there first. The

Cubs haven't been there since 1945 and have been to only four divisional series since, which doesn't give them many chances to win a championship. Were the Cubs consistently unlucky not winning their division? Did they just miss the divisional title a number of times because they were competing head to head with their very successful rivals the St. Louis Cardinals, whose ten World Series titles put them second behind the Yankees? If so, the Cubs should finish second far more often than they do third or fourth or last.

Alas, the Cubs have finished second even *fewer* times than first. They have finished third more times than first or second, finished fourth more times than third, and finished dead last 17 times. This evidence is not consistent with luck. Luck should have no order to it. Luck implies that you are equally likely to finish second as you are to do anything else. The Cubs' consistent placement toward the bottom is not a matter of luck. They have reached the bottom far more often than random chance says they should, finishing last or second to last nearly 40 percent of the time. The odds of this happening by chance are 527 to 1.

For comparison, the Yankees have been to 40 World Series (winning 27) and have finished first in their division (or league, in the early part of the twentieth century) 45 out of 100 times (the Cubs only 12), and when they haven't finished first, they've usually finished second (16 times). In fact, the Yankees' experience is opposite to that of the Cubs. The Yankees finish first (far) more often than second, finish second more often than third, finish third more often than fourth, and have finished dead last only three times. This is also not consistent with luck.

If you want to pity a team that is unlucky, consider the Houston Astros. That team has never won a World Series in its 48-year existence despite reaching the League Championship Series four times and the Division Series seven times. The Astros have also finished in the top three in their division 26 out of 48 years—more than 54 percent of the time—and have finished last in only three seasons.

Another way to measure luck is to see how much of a team's success or failure can't be explained. For example, take a look at how the team performed on the field and whether, based on its performance, it won fewer games than it should have. If you were told that your team led the league in hitting, home runs, runs scored, pitching, and fielding percentage, you'd assume your team won a lot more games than it lost. If it did not, you'd be within your rights to consider it unlucky. A lot would be left unexplained. How, for instance, did the 1982 Detroit Tigers finish fourth in their division, winning only 83 games and losing 79, despite placing eighth in the Majors in runs scored that season, seventh in team batting average, fourth in home runs, tenth in runs against, ninth in ERA, fifth in hits allowed, eighth in strikeouts against, and fourth in fewest errors?

Historically, for the average MLB team, its on-the-field statistics would predict its winning percentage year to year with 93 percent accuracy. That is, if you were to look only at a team's on-the-field numbers each season and rank it based on those numbers, 93 percent of the time you would get the same ranking as you would if you ranked it based on wins and losses. But 93 percent is not 100 percent. And for the 1982 Detroit Tigers, this was one of those "unlucky" years in which performance on the field simply did not translate into actual wins and losses. Lucky teams are those whose records are not justified by their on-the-field performances—in other words, there is a lot unexplained.

Based on this measure, how unlucky are the Cubs? Did the Cubs lose more games than they should have based on their performance at the plate, on the mound, and in the field? Is there something unexplained about their lack of success—like a curse?

Unfortunately (for us Cubs fans), no. The Cubs' record can be explained just as easily as those of the majority of teams in baseball. The Cubs' ritual underperformance in terms of wins is perfectly understandable when you examine their performance on the field. To put it more precisely, if we were to predict year to year the Cubs' winning percentage based on all available statistics, we would be able to explain 94 percent of it, which is higher than the league average. Here you could argue that the Cubs are actually *less* unlucky than the average team in baseball.

Who has been most affected by luck? Or, put differently, whose regular season record and postseason success are the hardest to explain? Life being heavy into irony, it's the Cubs' rivals, the St. Louis Cardinals. If you look at their performance on the field, you'd predict fewer wins than the Cardinals have achieved. Even more irritating to Cubs fans, it is also hard to explain how the Cardinals won ten World Series. The Dodgers have been to one more World Series than the Cardinals (18 to 17) but have won four fewer times. The Giants have been to the Fall Classic one time more than the Cardinals but have won four fewer championships.

But if bad luck—or a deficiency of good luck—isn't the answer, what *is* driving the Cubs' futility?

To traffic in the obvious, the reason the Cubs haven't won is that they haven't put particularly skilled teams on the field. You could start by picking apart personnel moves over the years. In 1964, the Cubs dealt a young outfielder, Lou Brock, to the rival St. Louis Cardinals for pitcher Ernie Broglio; this is generally considered by some (mostly in Chicago) the single worst trade in baseball history. Broglio would go 7–19 with the Cubs. Brock would retire as baseball's all-time leader in stolen bases and enter the Hall of Fame. The Cubs' trade of Dennis Eckersley, a future MVP, for three minor leaguers would rank up there, too. So would the decision to let a promising young pitcher, Greg Maddux, test the free agent market in 1992. Over the next decade, Maddux, as a member of the Atlanta Braves, would establish himself as the National League's dominant pitcher and a lock for the Hall of Fame. But all teams make trades that with the benefit of hindsight are boneheaded.

The bigger question is why the Cubs haven't put good teams together. We believe that the answer has to do with incentives. What fans are attributing to bad luck may be masking something more disturbing about the franchise.

Apart from the ineffable reasons—pride, competitiveness, honor—sports teams have an economic reason or incentive to do well. A more successful team generates more fans, which generates more revenue. Winning teams should attract more sellout crowds and trigger larger demand for sponsorships and local and national TV ratings and souvenir sales. Overall, winning should boost the brand name of the franchise, and all these things should increase the team's bottom line. The opposite would be true of losing teams. Think of this as a way for fans to reward a team's owners when the team performs well and punish them when it doesn't. This process aligns the incentives of fans with those of the owners, who gain financially by winning.

Sure, every team wants to win, but not equally. We don't often think about teams having different incentives to win, in part because knowing a team's incentive is difficult. But we can try to infer incentives by looking at data in new ways. For example, how does home game attendance respond to team performance? Home attendance is just one measure of a team's popularity and revenue, but it is correlated with others, such as sponsorship and souvenir sales. Imagine a team whose fans are so loyal or numb that winning or losing would not change attendance or the fan base. Compare this team with a team whose fans are very fickle and sensitive. Wouldn't the second team have a greater incentive to win? Failing to do so would be costly.

■ ■ ■

Calculating the response of home game attendance to season performance for every MLB team over the last century, we get a measure of how sensitive fans are to team success. If this number equals one, it means that when a team wins 10 percent more games, attendance rises by 10 percent—in other words, one for one. Greater than one means attendance rises by more than 10 percent (fans are more sensitive to performance), and less than one means fans are not as sensitive to performance, creating fewer incentives to win.

So, how do the Cubs stack up? It turns out that their attendance is the least sensitive to performance in all of baseball (see the graph below). The sensitivity of attendance per game to winning percentage for the Cubs is only 0.6, much less than one. The league average is one. If the Dallas Cowboys are America's Team, the Cubs are America's Teflon team.

Contrast these figures with those of the Yankees, where attendance sensitivity is 0.9, meaning that attendance moves almost one for one with winning percentage. You might think this is the case because New York fans are notoriously harsh, more willing to punish their teams for bad performance, or that Yankee tickets are so expensive that at those prices the team had better be good. Or perhaps the fans have been spoiled by all the success and have consuming expectations. So maybe a better comparison is to Chicago's other baseball team, the White Sox, who not only share a city with the Cubs but also play in a ballpark with roughly the same seating capacity. As it turns out, the White Sox fans' sensitivity to wins is more than twice that of the Cubs fans and one of the highest in the league.

The stark differences between the Cubs and the White Sox in terms of fan attendance can be seen clearly even over the last decade. The tables on page 245 list the wins and losses, rank in

ATTENDANCE ELASTICITY TO WINNING

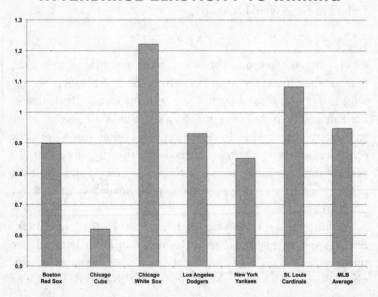

CHICAGO CUBS

SEASON	WINS	LOSSES	WIN %	RANK IN DIVISION	TOTAL ATTENDANCE	TOTAL CAPACITY	ATTEND%	+,- WIN%	+,- ATTEND%
1998	90	73	55%	2	2,623,194	3,189,964	82%		
1999	67	95	41%	6	2,813,854	3,151,062	89%	−14%	7%
2000	65	97	40%	6	2,789,511	3,151,062	89%	−1%	−1%
2001	88	74	54%	3	2,779,465	3,151,062	88%	14%	0%
2002	67	95	41%	5	2,693,096	3,034,356	89%	−13%	1%
2003	88	74	54%	1	2,962,630	3,151,062	94%	13%	5%
2004	89	73	55%	3	3,170,154	3,189,964	99%	1%	5%
2005	79	83	49%	4	3,099,992	3,151,062	98%	−6%	−1%
2006	66	96	41%	6	3,123,215	3,330,558	94%	−8%	−5%
2007	85	77	52%	1	3,252,462	3,330,558	98%	12%	4%
2008	97	64	60%	1	3,300,200	3,333,960	99%	8%	1%
2009	83	78	52%	2	3,168,859	3,333,960	95%	−9%	−4%

CHICAGO WHITE SOX

SEASON	WINS	LOSSES	WIN %	RANK IN DIVISION	TOTAL ATTENDANCE	TOTAL CAPACITY	ATTEND%	+,- WIN%	+,- ATTEND%
1998	80	82	49%	2	1,391,146	3,590,001	39%		
1999	75	86	47%	2	1,338,851	3,590,001	37%	−3%	−1%
2000	95	67	59%	1	1,947,799	3,590,001	54%	12%	17%
2001	83	79	51%	3	1,766,172	3,720,816	47%	−7%	−7%
2002	81	81	50%	2	1,676,911	3,720,816	45%	−1%	−2%
2003	86	76	53%	2	1,939,524	3,720,816	52%	3%	7%
2004	83	79	51%	2	1,930,537	3,186,216	61%	−2%	8%
2005*	99	63	61%	1	2,342,833	3,289,815	71%	10%	11%
2006	90	72	56%	3	2,957,414	3,289,815	90%	−6%	19%
2007	72	90	44%	4	2,684,395	3,289,815	82%	−11%	−8%
2008	89	74	55%	1	2,500,648	3,289,815	76%	10%	−6%
2009	79	83	49%	3	2,284,164	3,289,815	69%	−6%	−7%

*Won World Series.

division, and total attendance of the two Chicago teams from 1998 to 2009, including total seating capacity per season to account for stadium modifications.

Over the last 11 years, the rate of attendance at Wrigley Field wavered between 82 percent and 99 percent of capacity, whereas the White Sox had as little as 37 percent of their capacity filled in 1999 and as much as 90 percent of capacity filled in 2006—the year after they won the World Series, when attendance is always goosed. The same season, the Cubs would finish in last place, yet they posted a 94 percent attendance rate. The last-place Cubs entertained 165,801 more fans than the World Champion White Sox hosted that season! (And that doesn't include the thousands of rooftop seats enterprising landlords across the street from Wrigley rent out.) In fact, the Cubs have posted higher than 94 percent attendance rates in every season since 2002 despite not having even been to a World Series.

To see the general relation between winning and attendance, the last two columns of the table report the year-to-year percentage change in wins and the change in attendance relative to capacity. If attendance rises with winning—in other words, if fans create the right incentives—these numbers should move together. In 1999, the Cubs lost 14 percent more games than the previous year, yet attendance went *up* 7 percent. In 2001, the Cubs won 14 percent more games, yet attendance hardly budged. They lost 13 percent more games in 2002, yet attendance went *up* by 1 percent.

The precise opposite can be said of their South Side counterparts. White Sox attendance is, first of all, a lot lower than that of the Cubs in general. Only the year after winning the World Series did the White Sox hit the 90 percent attendance mark, and that faded to 69 percent within three years. Unlike the Cubs, year-to-year performance on the field directly translates into attendance for the Sox. In 2000, the White Sox won 12 percent more games and were rewarded with 17 percent more paying fans. In 2007, they lost 11 percent more games and were punished with 8 percent lower attendance. In almost every year, the change in attendance moves in the same direction as the team's performance. The Cubs' winning percentage can swing from year to year like Harry Caray's inning-to-inning blood-alcohol level used to. Attendance rates at Wrigley, however, are as steady as a surgeon's hands.

If you want an extreme stress test of the Cubs' attendance

"stickiness," you can examine what happened in 1994 and 1995 during and after the MLB strike. In 1994, the Cubs posted an 81 percent attendance rate, and the White Sox a respectable 72 percent rate. In 1995, the year after the strike, when baseball interest was at an all-time low, the Cubs still posted a 69 percent attendance rate, whereas the White Sox dropped to under 50 percent. That's more than a 30 percent drop in attendance for the White Sox and only a 15 percent drop for the Cubs. In New York, the Yankees had an unheard-of 41 percent attendance rate that same year. Within three years, both the Cubs and the Yankees were back to their normal attendance levels: the Yankees because of their performance—they won the World Series in 1996, made the postseason in 1997, and won the World Series again in 1998—and the Cubs . . . well, because they are the Cubs.

Yes, you say, but aren't there other factors? What about the fact that the Cubs play in a picturesque, idyllic old ballpark—all steel and brick and ivy-shrouded walls—whereas the White Sox's home has all the charm of its name, U.S. Cellular Field? Or that the Cubs reside in the gentrified North Side of Chicago, whereas the White Sox play in a dodgy South Side neighborhood? This might be a partial explanation, but even so it would still distort incentives. Consider other teams that play in venerable downtown parks. Boston has much higher attendance sensitivity to performance than do the Cubs (0.9). So do teams such as the San Francisco Giants and the Baltimore Orioles, which play in swank downtown venues (both have attendance-to-performance sensitivities of 1.15).

How else can we measure hidden incentives? Well, the Cubs, despite their futility, are still the fifth most valuable team in MLB according to *Forbes*, behind only the two New York teams, Boston, and the Los Angeles Dodgers. In 2007, before the global recession, the franchise was valued at $1 billion. In the summer of 2009, the Tribune Company sold 95 percent of the Cubs as well as Wrigley Field and a stake in a television network to the Ricketts family for roughly $900 million, one of the highest prices ever paid for a sports property. The deal was consummated days after the Cubs finished another uninspiring season, 83–78. Though these are all big-market teams, the White Sox, who play in the same big market as the Cubs, rank only fourteenth in value and are estimated to be worth roughly $250 million less than the Cubs.

In fact, the Cubs are the only team among the top ten most valuable franchises that do not have a recent championship. The Angels, Braves, Giants, Cardinals, and Phillies round out the top ten. Thanks largely to the fidelity of their fans—which also generates a lucrative television contract with WGN—the Cubs enjoy one of the highest market valuations in MLB without having to earn it on the field.

So, at least financially, the Cubs seem to have far less incentive to perform than do other teams—less than the Yankees and Red Sox do and certainly less than the White Sox. This doesn't mean the Cubs don't *want* to win, but it does mean that the Cubs have less of a financial incentive to win.

Winning or losing is often the result of a few small things that require extra effort to gain a competitive edge: going the extra step to sign the highly sought-after free agent, investing in a strong farm team with diligent scouting, monitoring talent, poring over statistics, even making players more comfortable. All can make a difference at the margin, and all are costly. When the benefits of making these investments are marginal at best, why undertake them? Would you work 10 percent harder at work if you got a 10 percent raise? Maybe. Would you work 10 percent harder for a 5 percent raise? Less likely. Professional sports are not immune to the power of incentives, either.

Although the players aren't armed with graphs, balance sheets, or statistics, they can sense this lack of urgency. It might mean taking an extra week of paid leave on the disabled list or arriving late to batting practice. It might mean missing a game, as Cubs outfielder José Cardenal once did, with the alibi "my eyelid [was] stuck open." It might mean antagonizing teammates in the clubhouse with earsplitting salsa music and then, made aware of their complaints, responding, "F— my teammates," as Sammy Sosa once did.

Joe Girardi was born in Peoria, rooted for the Cubs as a kid, and played seven seasons at Wrigley Field. He was devastated to leave the team. But then he played for the Yankees and won three World Series rings. Today he is the Yankees' manager. Girardi was once asked by *Harper's* magazine why the Cubs were so futile. He was quick to note that the Yankees' payroll was monstrous—and that didn't include "all the money spent on the minor leagues and free agents, signing kids from the Dominican, from Puerto Rico." That's it? "But it's more than that. In New York, you go into spring training expecting to get to the World Series. You feel it when you walk in the clubhouse—the pictures of all those Yankee greats, the monuments. There is something special about putting on the pinstripes. In Chicago, they hope for a good season, maybe the playoffs."

How deeply ingrained is this in the Cubs' culture? When P. K. Wrigley inherited the team from his father, the chewing gum tycoon William Wrigley, in 1932, he decided not to waste resources on baseball. According to *Harper's,* he decided that fans needed a reason, apart from the game, to venture to the ballpark. "The fun . . . the sunshine, the relaxation. Our idea is to get the public to go see a ball game, win or lose." To that end, he ordered one of his young employees, Bill Veeck, to plant ivy on the walls. Is it any wonder that more than 75 years later the team would still market the Wrigley experience, win or lose?

When they bought the team in 2009, the Ricketts family made capital improvements. One of the first moves of the new regime was to purge the clubhouse of ice cream, soda, and candy. They installed a stainless steel kitchen and hired a nutritional consultant. This was at the behest of Todd Ricketts, a fitness enthusiast who rightfully wondered whether a healthier, more energized team didn't stand a better chance of winning. The Cubs upgraded their scouting infrastructure, especially in Latin America, and entered the 2010 season with the highest payroll in the National League. As Tom Ricketts put it to us: "You obviously do it wisely, but you can't be afraid to spend when you think it will come back to benefit you on the field." He's right, of course. But the fact that this philosophy is such a marked departure from that of earlier ownerships goes a long way toward explaining the previous century of futility.

■ ■ ■

Some go so far as to argue that the Cubs may have had a perverse incentive to maintain their image as a "lovable loser," that the awfulness and the perpetual Charlie Brown status are part of the appeal of the franchise. Anger gave way to resignation years (decades?) ago. Fans sharing common failure become even bigger fans. Bonds fortify over "wait till next year." There is equity in futility.

Holly Swyers, an anthropology professor now at Lake Forest College in suburban Chicago, studied an indigenous tribe of Cubs fans that calls itself "the Regulars" (never, pointedly, the more common Bleacher Bums). The Regulars form a bona fide community, demonstrating a lot of the essential traits of a close-knit neighborhood, a church congregation, or even a family. The Regulars—a few hundred adults of mixed age, gender, and ethnicity, ranging from millionaire CEOs to retirees on fixed incomes—commune, rejoice and mourn together, and even marry among themselves. (Even Tom Ricketts met his future wife in the Wrigley bleachers.) Swyers noticed something else. The Regulars, who otherwise had little in common, also bonded over the team's misfortunes. They all are members of the same congregation, sharing pews in the Church of the Miserable. If the team were somehow to win a World Series, yes, the Regulars would be in nirvana. But the success would change the nature of the community.

Going back to financial incentives, it turns out that from 1990 to 2009 every team in MLB gained value the more it won—except one. The Cubs' franchise value actually rose slightly the more it lost! Why? Because fans kept coming despite the Cubs' ritual poor play. Gate revenue from ticket sales actually went up slightly when the Cubs lost a little more than usual, and TV revenue didn't change at all despite significant differences in winning percentages from season to season.

So if winning isn't what makes the Cubs valuable, what is it that keeps fans coming? If you've been to Wrigley Field in the last few decades, it's likely that you know part of the answer. Having spent a not insignificant part of his spring semesters at Purdue University going to games with his fraternity, one author of this book recalls vividly the drunkenness, the cute coeds, and the fun of evading authorities of all kinds. But there is no memory of games won or lost, of cheering, or of scorekeeping. Inquiries of "What's the score?" or "How many outs?" were abruptly met with showers of beer and slurred chants or drinking songs. As the popular T-shirt reads, "Cubs baseball: Shut up and drink your beer." A game at Wrigley is a party, maybe the best party in baseball (and don't forget that beer and concession sales generate revenue as well).

In 1983, in one of the finer sports monologues—a soliloquy that true die-hard Cubs fans can recite verbatim—Lee Elia, the Chicago manager at the time, alluded to this sensibility in graphic terms. After drunken bleacher bums booed the team during a desultory loss early in the season, Elia remarked in part: "What the f— am I supposed to do, go out there and let my f—ing players get destroyed every day and be quiet about it? For the f—ing nickel-dime people who turn up? The motherf—s don't even work. That's why they're out at the f—ing game. They oughta go out and get a f—ing job and find out what it's like to go out and earn a f—ing living. Eighty-five percent of the f—ing world is working. The other fifteen percent come out here."

Elia knew more than he thought he knew. Attendance at Wrigley is actually more sensitive to beer prices—much more—than it is to the Cubs' winning percentage. Obtaining beer prices from 1984 to 2009 and adjusting them for general price levels and inflation over this period, attendance was more than four times more sensitive to beer prices than to winning or losing.

What's more, the Cubs organization has understood this. Despite posting an abysmal 48.6 percent winning percentage over the last two decades, the Cubs' owners managed to increase ticket prices by 67 percent since 1990, which is way above the league average of 44.7 percent, *and* attendance still climbed to an all-time high 99 percent of capacity. But beer prices, not unlike the beer itself, remained pretty flat. By 2009, according to Team Marketing Report, Wrigley Field had the third-highest ticket prices in all MLB, averaging nearly $48 a seat, lagging behind only Fenway Park in Boston at $50 and the new Yankee Stadium at $73 a ticket. But the price of a small beer at Wrigley Field was the third *lowest* in the league ($5 at the concession stand, which is how TMR reports prices). Only the small-market Pittsburgh Pirates (at $4.75 a beer) and medium-market Arizona Diamondbacks (at $4.00) had cheaper beer—and their average ticket prices were $15.39 and $14.31, respectively.

In other words, Cubs fans will tolerate bad baseball *and* high ticket prices but draw the line at bad baseball and expensive beer. That makes for a fun day at the ballpark but doesn't give the ownership much incentive to reverse the culture of losing.

Oh, and just so you don't think this is simply a case of Chicagoans liking their baseball with (cheap) beer, White Sox attendance was unaffected by beer prices over the same period. White Sox fans, however, were more sensitive to ticket prices, and ticket prices tended to rise only after the team's winning percentage improved. So the White Sox understood their fans, too. In 2009, the average price of a ticket to see the White Sox was only $32, more than $15 cheaper than the price to see their North Side counterparts. But the same beer at U.S. Cellular Field would cost you $6.50—a 30 percent markup from the Wrigley vendors.

Bottom line: You'd be hard-pressed to call the Cubs' baseball (mis)fortunes a curse or to blame them on Steve Bartman, who, by the way, wasn't drinking that night, cheap beer or not.

EPILOGUE

If you had half as much fun reading this book as we had writing it, then we're all doing pretty well. Putting sports conventional wisdom to the test? Marrying sports with economic analysis, writing with statistics? Answering questions we've always pondered? Reconnecting with a childhood friend? According to our publisher, this was "work for hire," but, in truth, the contract should have read "play for hire."

There was, however, one serious problem. Even after submitting the manuscript to our editor, we had a hell of a time settling on a title. Like a pair of Mad Men—save the heroic drinking—we kicked ideas and concepts back and forth. We wanted a catchy phrase that captured both the sports component and the behavioral economics component. We were after something with both heft and levity. We didn't want to turn off the casual sports fan with jargon, but we also wanted to convey some rigor and sophistication. *Mathletes?* Too geeky. *Inside the Helmet?* Too trite. *Streakanomics?* Too derivative. *Why We Win?* Too self-helpy. *Unforced Errors?* Too negative. *Breaking Balls?* Too much potential for an unfortunate double entendre.

Unlike naming our kids, we couldn't delegate the task to our wives. So on it went. We'd disagree—one of us digging *We're #1*, the other having a gag reflex every time it was invoked. We'd love a title at first (*I Got It*) and then hate it an hour later. We'd come up with another but then sour when we realized it didn't lend itself to an arresting cover illustration.

Finally, we had the good sense to remember some of the principles we've espoused in this book. *There's value in data. The bigger the sample size, the more accurate the information. Personal biases and tastes can be mitigated when confronted with independent data. Considering new ways of looking at the problem can provide a new perspective that may help solve it.*

So we polled family members, friends, and colleagues. Why stop there? Next, we solicited title ideas from *Sports Illustrated* readers and Twitter followers. It wasn't just that we expanded our sample size. There was now real diversity, men in Canada submitting ideas one minute; women in India weighing in a moment later.

Much like our army of unpaid consultants, the ideas were all over the map. *Give Up Hope. By a Shoestring. Impure Luck. @#$% my Regression. Daddy, Why Does Sports Radio Lie to Me? Non-Fantasy Football.* Then there were the academic titles: *Data Analysis and Behavioral Psychology in Sports from an Economic Perspective.* Hostile titles: *I Bet Your Team Will Lose, Dumbass.* Even the religious: *God Wanted Us to Win.* In a nod to our passion for tennis, one reader suggested *Johan Kriek-onomics.* (There was also *Jimmy the Greek-onomics.*) In addition to seeking suggestions from the masses, we hired a consultant, who not only provided title suggestions but also helped us ferret out the best ones we'd received. (Incidentally, and not accidentally, the consultant was Linda Jines—who coined the phrase *Freakonomics* after its authors went through a similarly agonizing title search.)

So we gathered outside independent opinions. Lots of them. We may have known our book better than anyone, but we'd be fools to think we have all the answers and that we can't learn from a much wider set of ideas.

In the end, a mixture of data and expertise (thank you, Linda) helped us converge on the title. We finally settled on *Scorecasting* (credit to Jeff Boesiger for the original suggestion). As we've tried to emphasize throughout the book: Ignore data and diverse views at your peril. Seeking controversial or opposite opinions and challenging convention improve your decision-making. Book titles included. *Scorecasting* might not have been everyone's favorite title. But Lord knows, it had empirical backing.

In keeping with this theme, we'd like to solicit more ideas from you. For all the topics we explored, there were plenty of others we couldn't get to. At least not this time. But with any luck we'll write a sequel, and we suspect many of you have long-standing sports questions you'd like to put to the data. If so, we'd be happy to do the dirty work and test them. We are certain that, collectively, you will come up with intriguing ideas we hadn't considered. Send them to Scorecasting.com or check out the book's Facebook page.

ACKNOWLEDGMENTS

We pitched this book as a collaborative effort, but soon the collaboration went well beyond the two of us. Thanks are in order to Roger Scholl at Random House, a writer's (and, for that matter, economist's) editor who "got" the idea immediately and trusted us to deliver. Thanks also to Roger's colleagues Christine Kopprasch and production editor Mark Birkey.

Inasmuch as writing a book is likened to childbirth, Scott Waxman, our agent, was first a capable midwife and then a fine pediatrician. We owe a debt of gratitude to a small army who contributed ideas, comments, anecdotes, interviews, and stories for the book: Mike Carey, Frank Cheng, Tomago Collins, Joshua Coval, Mark Cuban, Jessica Dosen, Welington Dotel, David Epstein, Eugene Fama, Tom Gilovich, Jeff Heckelman, John Huizinga, Kevin Kelley, Steven Levitt, Cade Massey, Mike McCoy, Jack Moore, Daryl Morey, Mike Morin, Natxo Palacios-Huerta, Jeff Pearlman, Tom Perrotta, Devin Pope, Gregg Popovich, Tom Ricketts, Ryan Rodenberg, Scott Rosner, Jeff Spielberger, Susan Szeliga, Richard Thaler, Shino Tsurubuchi, and Charles "Sandy" Weil.

A special thank-you to Daniel Cervone, a University of Chicago undergrad so enamored with sports that he decided to spend a year between college and grad school helping us gather, organize, and analyze an absolutely massive amount of data. Good luck with your doctoral studies at Harvard—we suspect we'll be hearing great things from you soon.

Also, special thanks to Rebecca Sun, who challenged and improved our material with her sharp mind, sharp eye, and painstaking attention to detail. (Every author should inquire about her services.) So long as *Sports Illustrated* continues to attract talent of her caliber, it bodes well for the continued strength of the magazine.

PERSONAL ACKNOWLEDGMENTS

I owe a huge debt to my colleagues at the University of Chicago, my co-authors, and my thesis advisors from grad school at UCLA. Everything I've learned about economics and finance is owed to them. I am grateful to sit in the wonderful research environment at the Booth School of Business at the University of Chicago. There is simply no better place for novel research and critical thinking. Many thanks to my co-author, L. Jon Wertheim—whom I first knew as Lewis—who is not only a brilliant writer and unconventional thinker, but a great friend. It's been terrific catching up with him over the past two years. Finally, a heartfelt thanks to my wife, Bonnie, and to our children, Isaac, Josh, Sam, and little Sarah, who arrived right around chapter 10.

—TJM

I'm in arrears to Terry McDonell, Chris Hunt, and the other *Sports Illustrated* editors who could not have been more accommodating and supportive of this project. Warm thanks to my co-author, Tobias Moskowitz—whom I first knew as Toby—a first-rate thinker, economist, and analyst. He made a hell of a doubles partner on the Indiana junior tennis circuit, and a better one twenty years later on this project. Finally, my deepest thanks, as ever, are reserved for Ellie, Ben, and Allegra.

—LJW

BIBLIOGRAPHY

A list of relevant papers and articles organized by chapter, with web links where possible.

WHISTLE SWALLOWING

Asch, D. A., J. Baron, J. C. Hershey, H. Kunreuther, J. Meszaros, I. Ritov, and M. Spranca. "Omission Bias and Pertussis Vaccination." *Medical Decision Making* 14, no. 2 (Apr.–June 1994): 118–23.

Baron, J., G. B. Holzman, and J. Schulkin. "Attitudes of Obstetricians and Gynecologists Toward Hormone Replacement Therapy." *Medical Decision Making* 18, no. 4 (1998): 406–11.

Cervone, Daniel, and Tobias J. Moskowitz. "Whistle Swallowing: Officiating and the Omission Bias." Working paper, University of Chicago and Harvard University (Mar. 2010).

Kordes-de Vaal, H. Johanna. "Intention and the Omission Bias: Omissions Perceived as Nondecisions." *Acta Psychologica* 93 (1996): 161–72.

Price, Joseph, Marc Remer, and Daniel F. Stone. "Sub-Perfect Game: Profitable Biases of NBA Referees" (Dec. 1, 2009). Available at SSRN: *http://ssrn.com/abstract=1377964*

Spranca, M., E. Minsk, and J. Baron. "Omission and Commission in

Judgment and Choice." *Journal of Experimental Social Psychology* 27 (1991): 76–105.

GO FOR IT

Adams, Christopher. "Estimating the Value of 'Going for It' (When No One Does)," Dec. 12, 2006. *http://ssrn.com/abstract=950987*

Bisland, R. B. "A Stochastic CAI Model for Assisting in the Design of Football Strategy." *SIGSIM Simul. Dig.* 10, nos. 1–2 (Sep. 1978): 28–30. *http://doi.acm.org/10.1145/1102786.1102790*

Boronico, J. S., and S. L. Newbert. "Play Calling Strategy in American Football: A Game-Theoretic Stochastic Dynamic Programming Approach." *Journal of Sport Management* 13, no. 2 (1999): 103–13.

Carter, Virgil, and Robert E. Machol. "Optimal Strategies on Fourth Down." *Management Science* 24, no. 16 (Dec. 1978): 1758–62.

Romer, David H. "Do Firms Maximize?: Evidence from Professional Football." *Journal of Political Economy* (University of Chicago Press) 114, no. 2 (Apr. 2006): 340–65.

———. "It's Fourth Down and What Does the Bellman Equation Say?: A Dynamic Programming Analysis of Football Strategy." NBER Working Paper No. W9024 (June 2002). Available at SSRN: *http://ssrn.com/abstract=316803*

Rubenson, D. L. "On Creativity, Economics, and Baseball." *Creativity Research Journal* 4, no. 2 (1991): 205–9. *http://dx.doi.org/10.1080/10400419109534391*

HOW COMPETITIVE ARE COMPETITIVE SPORTS?

El-Hodiri, Mohamed, and James Quirk. "An Economic Model of a Professional Sports League." *Journal of Political Economy* (University of Chicago Press) 79, no. 6 (Nov.–Dec. 1971): 1302–19.

Humphreys, Brad R. "Alternative Measures of Competitive Balance in Sports Leagues." *Journal of Sports Economics* 3, no. 2 (May 2002): 133–48.

Neale, Walter C. "The Peculiar Economics of Professional Sports: A Contribution to the Theory of the Firm in Sporting Competition and in Market Competition." *The Quarterly Journal of Economics* 78, no. 1 (1964): 1–14.

Ross, Stephen F., and Stefan Szymanski. "Open Competition in League Sports." *Wisconsin Law Review,* 2002: 625. Available at SSRN: *http://ssrn.com/abstract=350960* or *http://dx.doi.org/10.2139/ssrn.350960*

Szymanski, Stefan. "The Champions League and the Coase Theorem." *Scottish Journal of Political Economy* 54, no. 3 (July 2007): 355–73. Available at SSRN: *http://ssrn.com/abstract=992044* or *http://dx.doi.org/10.1111/j.1467-9485.2007.00419.x*

Vrooman, John. "A General Theory of Professional Sports Leagues." *Southern Economic Journal* 61, no. 4 (Apr. 1995): 971–90.

TIGER WOODS IS HUMAN

Carmon, Ziv, and Dan Ariely. "Focusing on the Forgone: How Value Can Appear So Different to Buyers and Sellers." *Journal of Consumer Research*, 2000.

Kahneman, D., J. L. Knetsch, and R. H. Thaler. "Experimental Tests of the Endowment Effect and the Coase Theorem." *Journal of Political Economy* 98 (1990): 1325–48.

Kahneman, D., and A. Tversky. "Prospect Theory: An Analysis of Decisions Under Risk." *Econometrica* 47 (1979): 263–91.

Pope, Devin G., and Maurice E. Schweitzer. "Is Tiger Woods Loss Averse?: Persistent Bias in the Face of Experience, Competition, and High Stakes." *American Economic Review*, June 13, 2009.

Thaler, R. "Toward a Positive Theory of Consumer Choice." *Journal of Economic Behavior and Organization* 1 (1980): 39–60.

Tom, Sabrina M., Craig R. Fox, Christopher Trepel, and Russell A. Poldrack. "The Neural Basis of Loss Aversion in Decision-Making Under Risk." *Science* 315 (2007): 515–18.

THE VALUE OF A BLOCKED SHOT

Huizinga, J. "The Value of a Blocked Shot in the NBA: From Dwight Howard to Tim Duncan." MIT Sloan Sports Analytics Conference, 2010.

ROUNDING FIRST

Franke, R., W. Mayew, and Y. Sun. "Do Pennies Matter?: Investor Relations Consequences of Small Negative Earnings Surprises." *Review of Accounting Studies* 15, no. 1: 220–42.

Pope, D., and U. Simonsohn. "Round Numbers as Goals: Evidence from Baseball, SAT Takers, and the Lab." *Psychological Science*, in press.

THANKS, MR. ROONEY

Madden, Janice F., and Matthew Ruther. "Has the NFL's Rooney Rule Efforts 'Leveled the Field' for African American Head Coach Candidates?" Working paper, Wharton Business School, University of Pennsylvania, 2010.

SO, WHAT *IS* DRIVING THE HOME FIELD ADVANTAGE?

Asch, S. E. "Opinions and Social Pressure." *Scientific American* 193 (1955): 31–35.

Boyko, R., A. Boyko, and M. Boyko. "Referee Bias Contributes to Home Advantage in English Premiership Football." *Journal of Sports Sciences* 25, no. 11 (2007): 1185–94. *http://dx.doi.org/10.1080/ 02640410601038576*

Gandar, John M., Richard A. Zuber, and Reinhold P. Lamb. "The Home Field Advantage Revisited: A Search for the Bias in Other Sports Betting Markets." *Journal of Economics and Business* 53, no. 4 (Jul.–Aug. 2001): 439–53. ISSN 0148–6195, DOI 10.1016/ S0148–6195(01)00040–6 (*http://www.sciencedirect.com/science/ article/B6V7T-43DDWDC-6/2/120251992599a1f59f7edca7ee5f8be4*)

Johnston, R. "On Referee Bias, Crowd Size, and Home Advantage in the English Soccer Premiership." *Journal of Sports Sciences* 26, no. 6 (2008): 563–68. *http://dx.doi.org/10.1080/02640410701736780*

Nevill, A. M., N. J. Balmer, and A. Mark Williams. "The Influence of Crowd Noise and Experience upon Refereeing Decisions in Football." *Psychology of Sport and Exercise* 3, no. 4 (Oct. 2002): 261–72. ISSN 1469–0292; DOI 10.1016/S1469–0292(01)00033–4 (*http://www .sciencedirect.com/science/article/B6W6K-44B6RRV-1/2/0b8c6ecf53 aa433e8aeb7c55e9e2e2b4*)

Nevill, A. M., and R. L. Holder. "Home Advantage in Sport: An Overview of Studies on the Advantage of Playing at Home." *Sports Medicine* 28, no. 4 (Oct 1999): 221–36 (16).

Pettersson-Lidbom, Per, and Mikael Priks. "Behavior Under Social Pressure: Empty Italian Stadiums and Referee Bias." *Economics Letters* 108, no. 2 (Aug. 2010): 212–14. ISSN 0165–1765; DOI 10.1016/j .econlet.2010.04.023 (*http://www.sciencedirect.com/science/article/ B6V84-4YWYYNG-1/2/50f3eba6b28930d3ee61ac5d7093208d*)

Pollard, R. "Home Advantage in Soccer: A Retrospective Analysis." *Journal of Sports Sciences* 4, no. 3 (1986): 237–48. *http://dx.doi.org/10 .1080/02640418608732122*

Schwartz, Barry, and Stephen F. Barsky. "The Home Advantage." *Social Forces 55*, no. 3 (Mar. 1977): 641–62.

Sherif, Muzafer. *The Psychology of Social Norms.* New York: Harper Collins, 1936.

Smith, Erin E., and Jon D. Groetzinger. "Do Fans Matter?: The Effect of Attendance on the Outcomes of Major League Baseball Games." *Journal of Quantitative Analysis in Sports 6*, no. 1 (2010), article 4.

Sutter, Matthias, and Martin G. Kocher. "Favoritism of Agents: The Case of Referees' Home Bias." *Journal of Economic Psychology 25*, no. 4 (Aug. 2004): 461–69. ISSN 0167–4870; DOI: 10.1016/S0167–4870 (03)00013–8 (*http://www.sciencedirect.com/science/article/B6V8H -4841015–3/2/c440623bda16562409bc873e4592860f*)

Vergin, Roger C., and John J. Sosik. "No Place Like Home: An Examination of the Home Field Advantage in Gambling Strategies in NFL Football." *Journal of Economics and Business 51*, no. 1 (Jan. 2, 1999): 21–31. ISSN 0148–6195; DOI 10.1016/S0148–6195 (98)00025–3. (*http://www.sciencedirect.com/science/article/B6V7T -3XG7J46–2/2/27b182f8b90352382564cf7ce55344f8*)

OFF THE CHART

Eschker, E., S. J. Perez, and M. V. Siegler. "The NBA and the Influx of International Basketball Players." *Applied Economics 36*, no. 10 (2004): 1009–20. *http://dx.doi.org/10.1080/0003684042000246713*

Grier, Kevin B., and Robert D. Tollison. "The Rookie Draft and Competitive Balance: The Case of Professional Football." *Journal of Economic Behavior and Organization* (Elsevier) 25, no. 2 (Oct. 1994): 293–98.

Kahn, Lawrence M. "The Sports Business as a Labor Market Laboratory." *Journal of Economic Perspectives* (American Economic Association) 14, no. 3 (Summer 2000): 75–94.

Kahn, Lawrence M., and Malav Shah. "Race, Compensation, and Contract Length in the NBA: 2001–2002." *Industrial Relations* 44, no. 3 (July 2005): 444–62. Available at SSRN: *http://ssrn.com/ abstract=739935*

Massey, Cade, and Richard H. Thaler. "The Loser's Curse: Overconfidence vs. Market Efficiency in the National Football League Draft," Aug. 14, 2010. Available at SSRN: *http://ssrn.com/abstract=697121*

Rosen, Sherwin, and Allen Sanderson. "Labor Markets in Professional Sports." NBER Working Papers, National Bureau of Economic Research, Inc., no. 7573 (2000). *http://econpapers.repec.org/ RePEc:nbr:nberwo:7573*

Staw, Barry M., and Ha Hoang. "Sunk Costs in the NBA: Why Draft Order Affects Playing Time and Survival in Professional Basketball." *Administrative Science Quarterly* 40, no. 3 (Sep. 1995): 474–94.

Thaler, Richard. *The Winner's Curse: Paradoxes and Anomalies of Economic Life.* New York: Free Press, 1991.

HOW A COIN TOSS TRUMPS ALL

Burke, Brian, *advancedNFLstats.com*

Magnus, J. R., and F. J. G. M. Klaassen. "On the Advantage of Serving First in a Tennis Set: Four Years at Wimbledon." *The Statistician* (Journal of the Royal Statistical Society, Ser. D) 48 (1999): 247–56.

WHAT *ISN'T* IN THE MITCHELL REPORT?

Marcano, Arturo J., and David P. Fidler. "The Globalization of Baseball: Major League Baseball and the Mistreatment of Latin American Baseball Talent." *Indiana Journal of Global Legal Studies* 6, no. 2 (1999): 511.

Simpson, Tyler M. "Balking at Responsibility: Baseball's Performance-Enhancing Drug Problem in Latin America." *Law and Business Review of the Americas* 14 (2008): 369.

Spagnuolo, Diana L. "Swinging for the Fence: A Call for Institutional Reform as Dominican Boys Risk Their Futures for a Chance in Major League Baseball." *University of Pennsylvania Journal of International Economic Law* 24 (2003): 263.

DO ATHLETES REALLY MELT WHEN ICED?

Berry, S., and C. Wood. "The Cold-Foot Effect." *Chance* 17, no. 4 (2004): 47–51.

THE MYTH OF THE HOT HAND

Adams, R. M. "The 'Hot Hand' Revisited: Successful Basketball Shooting as a Function of Intershot Interval." *Perceptual and Motor Skills* 74 (1992): 934.

———. "Momentum in the Performance of Professional Tournament Pocket Billiards Players." *International Journal of Sport Psychology* 26 (1995): 580–87.

Albert, J. "A Statistical Analysis of Hitting Streaks in Baseball." *Journal of the American Statistical Association* 88 (1993): 1184–88.

Albert, Jim, and Jay Bennett. *Curve Ball: Baseball, Statistics, and the Role of Chance in the Game.* New York: Copernicus, 2001.

Albert, J., and P. Williamson. "Rejoinder." *Journal of the American Statistical Association* 88 (1993): 1194–96.

———. "Using Model/Data Simulations to Detect Streakiness." *The American Statistician* 55 (2001): 41–50.

Albright, S. C. "A Statistical Analysis of Hitting Streaks in Baseball." *Journal of the American Statistical Association* 88 (1993): 1175–83.

Ayton, P., and I. Fischer. "The Hot Hand Fallacy and the Gambler's Fallacy: Two Faces of Subjective Randomness?" *Memory and Cognition* 32 (2004): 1369–78.

Bar-Eli, Michael, Simcha Avugos, and Markus Raab. "Twenty Years of 'Hot Hand' Research: Review and Critique." *Psychology of Sport and Exercise* 7 (2006): 525–53.

Brown, W. O., and R. D. Sauer. "Does the Basketball Market Believe in the 'Hot Hand'?" *The American Economic Review* 83 (1993): 1377–86.

———. "Fundamentals or Noise? Evidence from the Professional Basketball Betting Market." *Journal of Finance* 48 (1993): 1193–1209.

Burns, B. D., and B. Corpus. "Randomness and Inductions from Streaks: 'Gambler's Fallacy' Versus 'Hot Hand.'" *Psychonomic Bulletin and Review* 11 (2004): 179–84.

Camerer, C. F. "Does the Basketball Market Believe in the 'Hot Hand'?" *The American Economic Review* 79 (1989): 1257–61.

Carhart, Mark M. "On Persistence in Mutual Fund Performance." *Journal of Finance* 52, no. 1 (Mar. 1997).

Clark, R. D. "An Analysis of Streaky Performance on the LPGA Tour." *Perceptual and Motor Skills* 97 (2003): 365–70.

———. "Examination of Hole-to-Hole Streakiness on the PGA Tour." *Perceptual and Motor Skills* 100 (2005): 806–14.

———. "Streakiness Among Professional Golfers: Fact or Fiction?" *International Journal of Sport Psychology* 34 (2003): 63–79.

Croucher, J. S. "An Analysis of the First 100 Years of Wimbledon Tennis Finals." *Teaching Statistics* 3 (1981): 72–75.

Dorsey-Palmateer, R., and G. Smith. "Bowlers' Hot Hands." *The American Statistician* 58 (2004): 38–45.

Forthofer, R. "Streak Shooter: The Sequel." *Chance* 4 (1991): 46–48.

Frame, D., E. Hughson, and J. C. Leach. "Runs, Regimes, and Rationality: The Hot Hand Strikes Back." Working paper, 2003.

Frohlich, C. "Baseball: Pitching No-Hitters." *Chance* 7 (1994): 24–30.

Gandar, John, Richard Zuber, Thomas O'Brien, and Ben Russo. "Testing Rationality in the Point Spread Betting Market." *Journal of Finance* 43 (1988): 995–1008.

Gilovich, T. "Judgmental Biases in the World of Sports." In W. F. Straub and J. M. Williams (eds.), *Cognitive Sport Psychology*. Lansing, New York: Sport Science Associates, 1984.

———. Transcript from Tom Gilovich hot hand homepage online chat, 2002. *http://www.hs.ttu.edu/hdfs3390/hh_gilovich.htm*

Gilovich, T., R. Vallone, and A. Tversky. "The Hot Hand in Basketball: On the Misperception of Random Sequences." *Cognitive Psychology* 17 (1985): 295–314.

Golec, J., and M. Tamarkin. "The Degree of Inefficiency in the Football Betting Markets." *Journal of Financial Economics* 30 (1991): 311–23.

Gould, S. J. "The Streak of Streaks." *Chance* 2 (1989): 10–16.

Gula, B., and M. Raab. "Hot Hand Belief and Hot Hand Behavior: A Comment on Koehler and Conley." *Journal of Sport and Exercise Psychology* 26 (2004): 167–70.

Hendricks, D., J. Patel, and R. Zeckhauser. "Hot Hands in Mutual Funds: Short-Run Persistence of Relative Performance, 1974–1988." *The Journal of Finance* 48 (1993): 93–130.

Jackson, D., and K. Mosurski. "Heavy Defeats in Tennis: Psychological Momentum or Random Effect?" *Chance* 10 (1997): 27–34.

Kahneman, D., P. Slovic, and A. Tversky. *Judgment Under Uncertainty: Heuristics and Biases*. New York: Cambridge University Press, 1982.

Kahneman, D., and A. Tversky. "On the Psychology of Prediction." *Psychological Review* 80 (1973): 237–51.

Klaassen, F. J. G. M., and J. R. Magnus. "Are Points in Tennis Independent and Identically Distributed?: Evidence from a Dynamic Binary Panel Data Model." *Journal of the American Statistical Association* 96 (2001): 500–509.

Koehler, J. J., and C. A. Conley. "The 'Hot Hand' Myth in Professional Basketball." *Journal of Sport and Exercise Psychology* 25 (2003): 253–59.

Krueger, Samuel. "Persistence in Mutual Fund Performance: Analysis of Holdings Returns." Working paper, University of Chicago, Booth School of Business, 2009.

Miyoshi, H. "Is the 'Hot Hands' Phenomenon a Misperception of Random Events?" *Japanese Psychological Research* 42 (2000): 128–33.

Raab, M. "Hot Hand in Sports: The Belief in Hot Hand of Spectators in Volleyball." In M. Koskolou, N. Geladas, and V. Klissouras (eds.),

ECSS proceedings, vol. 2. Seventh Congress of the European Congress of Sport Sciences. Athens: Trepoleos, 2002, p. 971.

Sauer, Raymond D., Vic Brajer, Stephen P. Ferris, and M. Wayne Marr. "Hold Your Bets: Another Look at the Efficiency of the Gambling Market for NFL Games." *Journal of Political Economy* 96 (1988): 206–13.

Stern, H. S., and C. N. Morris. "A Statistical Analysis of Hitting Streaks in Baseball." *Journal of the American Statistical Association* 88 (1993): 1189–94.

Tversky, A., and T. Gilovich. "The Cold Facts About the 'Hot Hand' in Basketball." *Chance* 2 (1989): 16–21.

———. "The 'Hot Hand': Statistical Reality or Cognitive Illusion?" *Chance* 2 (1989): 31–34.

Tversky, A., and D. Kahneman. "Availability: A Heuristic for Judging Frequency and Probability." *Cognitive Psychology* 5 (1973): 207–32.

———. "Belief in the Law of Small Numbers." *Psychological Bulletin* 76 (1971): 105–10.

———. "Judgment Under Uncertainty: Heuristics and Biases." *Science* 185 (1974): 1124–31. *http://www.sciencemag.org/cgi/content/short/185/4157/1124*

Vergin, R. C. "Winning Streaks in Sports and the Misperception of Momentum." *Journal of Sport Behavior* 23 (2000): 181–97.

Vergin, Roger C., and Michael Scriabin. "Winning Strategies for Wagering on National Football League Games." *Management Science* 24 (1978): 809–18.

Wardrop, R. L. "Simpson's Paradox and the Hot Hand in Basketball." *The American Statistician* 49 (1995): 24–28.

———. "Statistical Tests for the Hot-Hand in Basketball in a Controlled Setting" (1999). *http://www.stat.wisc.edu/_wardrop/papers/tr1007.pdf*

Woodland, Linda M., and Bill M. Woodland. "Market Efficiency and the Favorite-Longshot Bias: The Baseball Betting Market." *Journal of Finance* 49 (1994): 269–79.

Zuber, Richard A., John M. Gandar, and Benny D. Bowers. "Beating the Spread: Testing the Efficiency of the Gambling Market for National Football League Games." *Journal of Political Economy* 93 (1985): 800–806.

OTHER USEFUL RESOURCES

The Hardball Times: *http://www.hardballtimes.com*
APBRmetrics: *http://sonicscentral.com/apbrmetrics/viewforum.php?f=1*
Football Outsiders: *http://www.footballoutsiders.com*

INDEX

ABOUT THE AUTHORS

TOBIAS J. MOSKOWITZ is the Fama Family Professor of Finance at the University of Chicago. He is the winner of the 2007 Fischer Black Prize, which honors the top finance scholar in the world under the age of forty.

L. JON WERTHEIM is a senior writer for *Sports Illustrated,* a recent Ferris Professor at Princeton, and the author of five books, including *Strokes of Genius: Federer, Nadal, and the Greatest Match Ever Played.*

ALSO BY L. JON WERTHEIM AND SAM SOMMERS

"Eye-opening, captivating, and hilarious, *This Is Your Brain on Sports* shines a fascinating and scientific spotlight on human nature."
—Amy Cuddy, Harvard Business School professor and author of *Presence*

THREE RIVERS PRESS
NEW YORK

AVAILABLE WHEREVER BOOKS ARE SOLD